A Quick Guide to API 510 Certified Pressure Vessel Inspector Syllabus

QG Publishing is a Matthews Engineering Training Ltd company

MATTHEWS
ENGINEERING TRAINING LTD
www.matthews-training.co.uk

Training courses for industry

- Plant in-service inspection training
- Pressure systems/PSSR/PED/PRVs
- Notified Body training
- Pressure equipment code design ASME/BS/EN
- API inspector training (UK) : API 510/570/653
- On-line training courses available

Matthews Engineering Training Ltd provides training in pressure equipment and inspection-related subjects, and the implementation of published codes and standards.

More than 500 classroom and hands-on courses have been presented to major clients from the power, process, petrochemical and oil/gas industries.

We specialize in in-company courses, tailored to the needs of individual clients.

Contact us at enquiries@matthews-training.co.uk
Tel: +44(0) 7732 799351

Matthews Engineering Training Ltd is an Authorized Global Training provider to The American Society of Mechanical Engineers (ASME)

www.matthews-training.co.uk

A Quick Guide to

API 510 Certified Pressure Vessel Inspector Syllabus

Example Questions and Worked Answers

Clifford Matthews

Series editor: Clifford Matthews

Matthews Engineering Training Limited
www.matthews-training.co.uk

WOODHEAD PUBLISHING LIMITED

Oxford   Cambridge   New Delhi

Published by Woodhead Publishing Limited, Abington Hall, Granta Park,
Great Abington, Cambridge CB21 6AH, UK
www.woodheadpublishing.com
and
Matthews Engineering Training Limited
www.matthews-training.co.uk

Woodhead Publishing India Private Limited, G-2, Vardaan House,
7/28 Ansari Road, Daryaganj, New Delhi – 110002, India

Published in North America by the American Society of Mechanical Engineers
(ASME), Three Park Avenue, New York, NY 10016-5990, USA
www.asme.org

First published 2010, Woodhead Publishing Limited and Matthews
Engineering Training Limited
© 2010, C. Matthews
The author has asserted his moral rights.

This book contains information obtained from authentic and highly regarded sources. Reprinted material is quoted with permission, and sources are indicated. Reasonable efforts have been made to publish reliable data and information, but the author and the publishers cannot assume responsibility for the validity of all materials. Neither the author nor the publishers, nor anyone else associated with this publication, shall be liable for any loss, damage or liability directly or indirectly caused or alleged to be caused by this book.

Neither this book nor any part may be reproduced or transmitted in any form or by any means, electronic or mechanical, including photocopying, microfilming and recording, or by any information storage or retrieval system, without permission in writing from Woodhead Publishing Limited.

The consent of Woodhead Publishing Limited does not extend to copying for general distribution, for promotion, for creating new works, or for resale. Specific permission must be obtained in writing from Woodhead Publishing Limited for such copying.

Trademark notice: Product or corporate names may be trademarks or registered trademarks, and are used only for identification and explanation, without intent to infringe.

British Library Cataloguing in Publication Data
A catalogue record for this book is available from the British Library.

Library of Congress Cataloging in Publication Data
A catalog record for this book is available from the Library of Congress.

Woodhead Publishing ISBN 978-1-84569-755-6 (print)
Woodhead Publishing ISBN 978-0-85709-102-4 (online)
ASME ISBN 978-0-7918-5962-9
ASME Order No. 859629
ASME Order No. 85962Q (e-book)

Typeset by Data Standards Ltd, Frome, Somerset, UK
Printed in the United Kingdom by Henry Ling Limited

Contents

The Quick Guide Series — x
How to Use This Book — xii

Chapter 1: Interpreting ASME and API Codes
1.1 Codes and the real world — 1
1.2 ASME construction codes — 1
1.3 API inspection codes — 2
1.4 Code revisions — 5
1.5 Code illustrations — 5
1.6 New construction versus repair activity — 6
1.7 Conclusion: interpreting API and ASME codes — 7

Chapter 2: An Introduction to API 510 (Sections 1–4)
2.1 Introduction — 10
2.2 Section 1: scope — 10
2.3 Section 3: definitions — 13
2.4 Section 4: owner/user/inspector organization — 16
2.5 API 510 sections 1–4 familiarization questions — 20

Chapter 3: API 510 Inspection Practices (Section 5)
3.1 Introduction to API 510 section 5: inspection practices — 23
3.2 Inspection types and planning — 23
3.3 Condition monitoring locations (CMLs) — 27
3.4 Section 5.8: pressure testing — 27
3.5 API 510 section 5 familiarization questions — 30

Chapter 4: API 510 Frequency and Data Evaluation (Sections 6 and 7)
4.1 Introduction — 33
4.2 The contents of section 6 — 34
4.3 API 510 section 6 familiarization questions — 40
4.4 Section 7: inspection data evaluation, analysis and recording — 42
4.5 API 510 section 7 familiarization questions — 56

Chapter 5: API 510 Repair, Alteration, Re-rating (Section 8)
- 5.1 Definitions — 61
- 5.2 Re-rating — 62
- 5.3 Repairs — 68
- 5.4 API 510 section 8 familiarization questions — 72

Chapter 6: API 572 Inspection of Pressure Vessels
- 6.1 API 572 introduction — 75
- 6.2 API 572 section 4: types of pressure vessels — 77
- 6.3 API 572 section 4.3: materials of construction — 78
- 6.4 API 572 sections 5, 6 and 7 — 80
- 6.5 API 572 section 8: corrosion mechanisms — 80
- 6.6 API 572 section 9: frequency and time of inspection — 82
- 6.7 API 572 section 10: inspection methods and limitations — 82
- 6.8 API 572 section 10 familiarization questions — 84

Chapter 7: API 571 Damage Mechanisms
- 7.1 API 571 introduction — 89
- 7.2 The first group of DMs — 93
- 7.3 API 571 familiarization questions (set 1) — 94
- 7.4 The second group of DMs — 97
- 7.5 API 571 familiarization questions (set 2) — 98
- 7.6 The third group of DMs — 99

Chapter 8: API 576 Inspection of Pressure-Relieving Devices
- 8.1 Introduction to API 576 — 105
- 8.2 API 576 sections 3 and 4: types (definitions) of pressure-relieving devices — 106
- 8.3 Types of pressure-relieving device — 108
- 8.4 API 576 section 5: causes of improper performance — 115
- 8.5 API 576 section 6: inspection and testing — 119
- 8.6 API 576 familiarization questions — 120

Chapter 9: ASME VIII Pressure Design
- 9.1 The role of ASME VIII in the API 510 syllabus — 124
- 9.2 How much of ASME VIII is in the API 510 syllabus? — 125

Contents

9.3	ASME VIII clause numbering	128
9.4	Shell calculations: internal pressure	130
9.5	Head calculations: internal pressure	134
9.6	Set 1: shells/heads under internal pressure familiarization questions	141
9.7	ASME VIII: MAWP and pressure testing	143
9.8	Set 2: MAWP and pressure testing familiarization questions	152
9.9	External pressure shell calculations	155
9.10	Set 3: vessels under external pressure familiarization questions	160
9.11	Nozzle design	161

Chapter 10: ASME VIII Welding and NDE

10.1	Introduction	171
10.2	Sections UW-1 to UW-5: about joint design	171
10.3	UW-12: joint efficiencies	175
10.4	UW-11: RT and UT examinations	177
10.5	UW-9: design of welded joints	179
10.6	ASME VIII section UW-11 familiarization questions (set 1)	180
10.7	Welding requirements of ASME VIII section UW-16	181
10.8	ASME VIII section UW-16 familiarization questions (set 2)	183
10.9	RT requirements of ASME VIII sections UW-51 and UW-52	185
10.10	ASME VIII section UW-51/52 familiarization questions (set 3)	192

Chapter 11: ASME VIII and API 510 Heat Treatment

11.1	ASME requirements for PWHT	195
11.2	What is in UCS-56?	195
11.3	API 510 PWHT overrides	197
11.4	ASME VIII sections UCS-56 and UW-40: PWHT familiarization questions	201

Chapter 12: Impact Testing

12.1	Avoiding brittle fracture	203
12.2	Impact exemption UCS-66	204

12.3 ASME VIII section UCS-66: impact test exemption familiarization questions	211

Chapter 13: Introduction to Welding/API 577

13.1 Module introduction	212
13.2 Welding processes	212
13.3 Welding consumables	215
13.4 Welding process familiarization questions	220
13.5 Welding consumables familiarization questions	222

Chapter 14: Welding Qualifications and ASME IX

14.1 Module introduction	225
14.2 Formulating the qualification requirements	225
14.3 Welding documentation reviews: the exam questions	231
14.4 ASME IX article I	233
14.5 Section QW-140 types and purposes of tests and examinations	235
14.6 ASME IX article II	236
14.7 ASME IX articles I and II familiarization questions	237
14.8 ASME IX article III	239
14.9 ASME IX article IV	240
14.10 ASME IX articles III and IV familiarization questions	243
14.11 The ASME IX review methodology	245
14.12 ASME IX WPS/PQR review: worked example	247

Chapter 15: The NDE Requirements of API 510 and API 577

15.1 NDE requirements of API 510 and API 577	257
15.2 API 510 NDE requirements	257

Chapter 16: NDE Requirements of ASME V

16.1 Introduction	262
16.2 ASME V article 1: general requirements	262
16.3 ASME V article 2: radiographic examination	263
16.4 ASME V article 6: penetrant testing (PT)	269
16.5 ASME V articles 1, 2 and 6: familiarization questions	273
16.6 ASME V article 7: magnetic testing (MT)	274

16.7 ASME V article 23: ultrasonic thickness checking	277
16.8 ASME V articles 7 and 23 familiarization questions	280

Chapter 17: Thirty Open-Book Sample Questions

Chapter 18: Answers

18.1 Familiarization answers	292
18.2 Open-book sample questions answers	297

Chapter 19: The Final Word on Exam Questions

19.1 All exams anywhere: some statistical nuts and bolts for non-mathematicians	301
19.2 Exam questions and the three principles of whatever (the universal conundrum of randomness versus balance)	303
19.3 Exam selectivity	305

Appendix

Publications effectivity sheet for API 510 Exam Administration: 2 June 2010	309

Index
312

The Quick Guide Series

The *Quick Guide* data books are intended as simplified, easily accessed references to a range of technical subjects. The initial books in the series were published by The Institution of Mechanical Engineers (Professional Engineering Publishing Ltd), written by the series editor Cliff Matthews. The series is now being extended to cover an increasing range of technical subjects by Matthews Engineering Training Ltd.

The concept of the Matthews *Quick Guides* is to provide condensed technical information on complex technical subjects in a pocket book format. Coverage includes the various regulations, codes and standards relevant to the subject. These can be difficult to understand in their full form, so the *Quick Guides* try to pick out the key points and explain them in straightforward terms. This of course means that each guide can only cover the main points of its subject – it is not always possible to explain everything in great depth. For this reason, the *Quick Guides* should only be taken as that – a quick guide – rather than a detailed treatise on the subject.

Where subject matter has statutory significance, e.g. statutory regulation and reference technical codes and standards, then these guides do not claim to be a full interpretation of the statutory requirements. In reality, even regulations themselves do not really have this full status – many points can only be interpreted in a court of law. The objective of the *Quick Guides* is therefore to provide information that will add to the clarity of the picture rather than produce new subject matter or interpretations that will confuse you even further.

If you have any comments on this book, or you have any suggestions for other books you would like to see in the

Quick Guides series, contact us through our website: www.matthews-training.co.uk

Special thanks are due to Helen Hughes for her diligent work in typing the manuscript for this book.

Cliff Matthews
Series Editor

How to Use This Book

This book is a 'Quick Guide' to the API 510 Certified Pressure Vessel Inspector examination syllabus (formally called the 'body of knowledge' by API). It is intended to be of use to readers who:

- intend to study and sit for the formal API 510 Individual Certification Program (ICP) examination or
- have a general interest in the content of API 510 and its associated API/ASME codes, as they are applied to the in-service inspection of pressure vessels.

The book covers all the codes listed in the API 510 syllabus (the so-called 'effectivity list') but only the content that is covered in the body of knowledge. Note that in some cases (e.g. ASME VIII) this represents only a small percentage of the full code content. In addition, the content of individual chapters of this book is chosen to reflect those topics that crop up frequently in the API 510 ICP examination. Surprisingly, some long-standing parts of the API 510 body of knowledge have appeared very infrequently, or not at all, in recent examinations.

While this book is intended to be useful as a summary, remember that it cannot be a full replacement for a programme of study of the necessary codes. The book does not cover the entire API 510 ICP syllabus, but you should find it useful as a pre-training course study guide or as pre-examination revision following a training course itself. It is very difficult, perhaps almost impossible, to learn enough to pass the exam using only individual reading of this book.

This quick guide is structured into chapters – each addressing separate parts of the API 510 ICP syllabus. A central idea of the chapters is that they contain self-test questions to help you understand the content of the codes. These are as important as the chapter text itself – it is a well-proven fact that you retain more information by actively

How to Use This Book

searching (either mentally or physically) for an answer to a question than by the more passive activity of simply reading through passages or tables of text.

Most of the chapters can stand alone as summaries of individual codes, with the exception of the mock examination questions that contain cumulative content from all of the previous chapters. It therefore makes sense to leave these until last.

Code references dates

The API 510 ICP programme runs several times per year with examinations held in June, September and December. Each examination sitting is considered as a separate event with the examination content being linked to a pre-published code 'effectivity list' and body of knowledge. While the body of knowledge does not change much, the effectivity list is continually updated as new addenda or editions of each code come into play. Note that a code edition normally only enters the API 510 effectivity list twelve months after it has been issued. This allows time for any major errors to be found and corrected.

In writing this *Quick Guide* it has been necessary to set a reference date for the code editions used. We have used the effectivity list for the June 2010 examinations. Hence all the references used to specific code sections and clauses will refer to the code editions/revisions mentioned in that effectivity list. A summary of these is provided in the Appendix.

In many cases the numbering of code clauses remains unchanged over many code revisions, so this book should be of some use for several years into the future. There are subtle differences in the way that API and ASME, as separate organizations, change the organization of their clause numbering systems to incorporate technical updates and changes as they occur – but they are hardly worth worrying about.

Important note: the role of API

API have not sponsored, participated or been involved in the compilation of this book in any way. API do not issue past ICP examination papers or details of their question banks to any training provider, anywhere.

API codes are published documents, which anyone is allowed to interpret in any way they wish. Our interpretations in this book are built up from a record of running successful API 570/510/653 training programmes in which we have achieved a first-time pass rate of 85–90 % +. It is worth noting that most training providers either do not know what their delegates' pass rate is or don't publish it if they do. API sometimes publish pass rate statistics – check their website www.api.org and see if they do, and what they are.

Chapter 1

Interpreting ASME and API Codes

Passing the API ICP examination is, unfortunately, all about interpreting codes. As with any other written form of words, codes are open to interpretation. To complicate the issue, different forms of interpretation exist between code types; API and ASME are separate organizations so their codes are structured differently, and written in quite different styles.

1.1 Codes and the real world

Both API and ASME codes are meant to apply to the real world, but in significantly different ways. The difficulty comes when, in using these codes in the context of the API ICP examinations, it is necessary to distil both approaches down to a single style of ICP examination question (always of multiple choice, single-answer format).

1.2 ASME construction codes

ASME construction codes (VIII, V and IX) represent the art of the possible, rather than the ultimate in fitness for service (FFS) criteria or technical perfection. They share the common feature that they are written entirely from a new construction viewpoint and hence are relevant up to the point of handover or putting into use of a piece of equipment. Strictly, they are not written with in-service inspection or repair in mind. This linking with the restricted activity of new construction means that these codes can be prescriptive, sharp-edged and in most cases fairly definitive about the technical requirements that they set. It is difficult to agree that their content is not black and white, even if you do not agree with the technical requirements or acceptance criteria, etc., that they impose.

Do not make the mistake of confusing the definitive requirements of construction codes as being the formal arbiter of FFS. It is technically possible, in fact commonplace, to use an item safely that is outside code requirements as long as its integrity is demonstrated by a recognized FFS assessment method.

1.3 API inspection codes

API inspection codes (e.g. API 510) and their supporting recommended practice documents (e.g. API RP 572 and 576) are very different. They are not construction codes and so do not share the prescriptive and 'black and white' approach of construction codes.

There are three reasons for this:

- They are based around accumulated expertise from a *wide variety* of equipment applications and situations.
- The technical areas that they address (corrosion, equipment lifetimes, etc.) can be diverse and uncertain.
- They deal with technical *opinion*, as well as fact.

Taken together, these make for technical documents that are more of a technical way of looking at the world than a solution, unique or otherwise, to a technical problem. In such a situation you can expect *opinion* to predominate.

Like other trade associations and institutions, API (and ASME) operate using a structure of technical committees. It is committees that decide the scope of codes, call for content, review submissions and review the pros and cons of what should be included in their content. It follows therefore that the content and flavour of the finalized code documents are the product of committees. The output of committees is no secret – they produce fairly well-informed opinion based on an accumulation of experience, tempered, so as not to appear too opinionated or controversial, by having the technical edges taken off. Within these constraints there is no doubt that API codes do provide sound and fairly balanced technical opinion. Do not be surprised, however, if this opinion does not necessarily match your own.

1.3.1 Terminology

API and ASME documents use terminology that occasionally differs from that used in European and other codes. Non-destructive examination (NDE), for example, is normally referred to as non-destructive testing (NDT) in Europe and API work on the concept that an operative who performs NDE is known as the *examiner* rather than by the term *technician* used in other countries. Most of the differences are not particularly significant in a technical sense – they just take a little getting used to.

In some cases, meanings can differ *between* ASME and API codes (pressure and leak testing are two examples). API codes benefit from their principle of having a separate section (see API 510 section 3) containing definitions. These definitions are selective rather than complete (try and find an accurate explanation of the difference between the terms *approve* and *authorize*, for example).

Questions from the ICP examination papers are based solely on the terminology and definitions understood by the referenced codes. That is the end of the matter.

1.3.2 Calculations

Historically, both API and ASME codes were based on the United States Customary System (USCS) family of units. There are practical differences between this and the European SI system of units.

SI is a consistent system of units, in which equations are expressed using a combination of *base* units. For example:

$$\text{Stress}(S) = \frac{\text{pressure}(p) \times \text{diameter}(d)}{2 \times \text{thickness}(t)}$$

In SI units all the parameters would be stated in their base units, i.e.

Stress: N/m^2 (Pa)
Pressure: N/m^2 (Pa)
Diameter: m
Thickness: m

Compare this with the USCS system in which parameters may be expressed in several different 'base' units, combined with a multiplying factor. For example the equation for determining the minimum allowable corroded shell thickness of storage tanks is:

$$t_{min} = \frac{2.6(H-1)DG}{SE}$$

where t_{min} is in inches, fill height (H) is in feet, tank diameter (D) is in feet, G is specific gravity, S is allowable stress and E is joint efficiency.

Note how, instead of stating dimensions in a single base unit (e.g. inches) the dimensions are stated in the most convenient dimension for measurement, i.e. shell thickness in inches and tank diameter and fill height in feet. Remember that:

- This gives the same answer; the difference is simply in the method of expression.
- In many cases this can be easier to use than the more rigorous SI system – it avoids awkward exponential (10^6, 10^{-6}, etc.) factors that have to be written in and subsequently cancelled out.
- The written terms tend to be smaller and more convenient.

1.3.3 Trends in code units

Until fairly recently, ASME and API codes were written exclusively in USCS units. The trend is increasing, however, to develop them to express all units in dual terms USCS(SI), i.e. the USCS term followed by the SI term in brackets. Note the results of this trend:

- Not all codes have been converted at once; there is an inevitable process of progressive change.
- ASME and API, being different organizations, will inevitably introduce their changes at different rates, as their codes are revised and updated to their own schedules.
- Unit conversions bring with them the problem of *rounding*

errors. The USCS system, unlike the SI system, has never adapted well to a consistent system of rounding (e.g. to one, two or three significant figures) so errors do creep in.

The results of all these is a small but significant effect on the form of examination questions used in the ICP examination and a few more opportunities for errors of expression, calculation and rounding to creep in. On balance, ICP examination questions seem to respond better to being treated using pure USCS units (for which they were intended). They do not respond particularly well to SI units, which can cause problems with conversion factors and rounding errors.

1.4 Code revisions
Both API and ASME review and amend their codes on a regular basis. There are various differences in their approach but the basic idea is that a code undergoes several addenda additions to the existing edition, before being reissued as a new edition. Timescales vary – some change regularly and others hardly at all.

Owing to the complexity of the interlinking and cross-referencing between codes (particularly referencing *from* API *to* ASME codes) occasional mismatches may exist temporarily. Mismatches are usually minor and unlikely to cause any problems in interpreting the codes.

It is rare that code revisions are very dramatic; think of them more as a general process of updating and correction. On occasion, fundamental changes are made to material allowable stresses (specified in ASME II-D), as a result of experience with material test results, failures or advances in manufacturing processes.

1.5 Code illustrations
The philosophy on figures and illustrations differs significantly between ASME and API codes as follows:

- **ASME codes** (e.g. **ASME VIII**), being construction-based,

contain numerous engineering-drawing style figures and tables. Their content is designed to be precise, leading to clear engineering interpretation.
- **API codes** are not heavily illustrated, relying more on text. Both API 510 and its partner pipework inspection code, API 570, contain only a handful of illustrations between them.
- **API Recommended Practice (RP) documents** are better illustrated than their associated API codes but tend to be less formal and rigorous in their approach. This makes sense, as they are intended to be used as technical information documents rather than strict codes, as such. API RP 572 is a typical example containing photographs, tables and drawings (sketch format) of a fairly general nature. In some cases this can actually make RP documents more practically *useful* than codes.

1.6 New construction versus repair activity

This is one of the more difficult areas to understand when dealing with ASME and API codes. The difficulty comes from the fact that, although ASME VIII was written exclusively from the viewpoint of new construction, it is referred to by API 510 in the context of in-service *repair* and, to a lesser extent, *re-rating*. The ground rules (set by API) to manage this potential contradiction are as follows (see Fig 1.1).

- For new construction, ASME VIII is used – and API 510 plays no part.
- For repair, API 510 is the 'driving' code. In areas where it references 'the construction codes' (e.g. ASME VIII), this is followed *when it can be* (because API 510 has no content that contradicts it).
- For repair activities where API 510 and ASME VIII contradict, then API 510 takes priority. Remember that these contradictions are to some extent false – they only exist because API 510 is dealing with on-site repairs, while

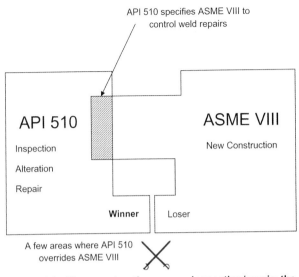

Figure 1.1 New construction versus inspection/repair: the ground rules

ASME VIII was not written with that in mind. Two areas where this is an issue are:
- some types of repair weld specification (material, fillet size, electrode size, etc.);
- how and when vessels are pressure tested.

1.7 Conclusion: interpreting API and ASME codes

In summary, then, the API and ASME set of codes are a fairly comprehensive technical resource, with direct application to plant and equipment used in the petroleum industry. They are perhaps far from perfect but, in reality, are much more comprehensive and technically consistent than many others. Most national trade associations and institutions do not have any in-service inspection codes *at all*, so industry has to rely on a fragmented collection from overseas sources or nothing at all.

The API ICP scheme relies on these ASME and API codes for its selection of subject matter (the so-called 'body of knowledge'), multiple exam questions and their answers. One of the difficulties is shoe-horning the different approach and style of the ASME codes (V, VIII and IX) into the same style of questions and answers that fall out of the relevant API documents (in the case of the API 510 ICP these are API 571/

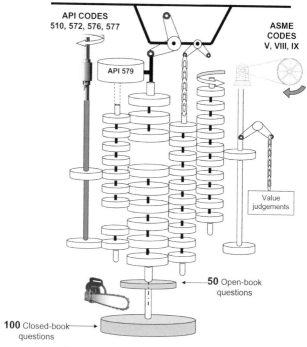

Figure 1.2　Codes in, questions out

572/576/577). Figure 1.2 shows the situations. It reads differently, of course, depending on whether you are looking for reasons for difference or seeking some justification for similarity. You can see the effect of this in the style of many of the examination questions and their 'correct' answers.

Difficulties apart, there is no question that the API ICP examinations are all about understanding and interpreting the relevant ASME and API codes. Remember, again, that while these codes are based on engineering experience, do not expect that this experience necessarily has to coincide with your own. Accumulated experience is incredibly wide and complex, and yours is only a small part of it.

Chapter 2

An Introduction to API 510 (Sections 1–4)

2.1 Introduction

This chapter is about learning to become familiar with the layout and contents of API 510. It forms a vital preliminary stage that will ultimately help you understand not only the content of API 510 but also its cross-references to the other relevant API and ASME codes.

API 510 is divided into nine sections (sections 1 to 9), five appendices (appendices A to E), one figure and two tables. Even when taken together, these are not sufficient to specify fully a methodology for the inspection, repair and re-rating of pressure vessels. To accomplish this, further information and guidance has to be drawn from other codes.

So that we can start to build up your familiarity with API 510, we are going to look at some of the definitions that form its basis. We can start to identify these by looking at the API 510 contents/index page. This is laid out broadly as shown in Fig. 2.1.

2.2 Section 1: scope

This is a very short (one-page) part of the code. The main point is in section 1.1.1, which states that all refining and chemical process vessels are included in the scope of API 510 except those vessels that are specifically *excluded* from the coverage of API 510. Note that this list (look at section 1.2.2) links together with a longer list in appendix A (look near the back of the document). Essentially, vessels that are excluded from the coverage of API 510 are:

- Mobile plant
- Anything designed to other parts of ASME
- Fired heaters

An Introduction to API 510

THE CONTENTS OF API 510: 9th EDITION

1. SCOPE
   1.1 General application
   1.2 Specific applications
   1.3 Recognized technical concepts

2. REFERENCES

3. DEFINITIONS

4. OWNER/USER INSPECTION ORGANIZATION
   4.1 General
   4.2 Owner–User organization responsibilities

5. INSPECTION PRACTICES
   5.1 Inspection plans
   5.2 RBI
   5.3 Preparation for inspection
   5.4 Inspection for damage mechanisms
   5.5 General inspection and surveillance
   5.6 Condition monitoring locations
   5.7 Condition monitoring methods
   5.8 Pressure testing
   5.9 Material verification and traceability
   5.10 Inspection of in-service welds and joints
   5.11 Inspection of flanged joints

6. INTERVAL/FREQUENCY AND EXTENT OF INSPECTION
   6.1 General
   6.2 Inspection during installation and service changes
   6.3 RBI
   6.4 External inspection
   6.5 Internal and on-stream inspection
   6.6 PRVs

7. INSPECTION DATA EVALUATION, ANALYSIS AND RECORDING
   7.1 Corrosion rate determination
   7.2 Remaining life calculations
   7.3 MAWP
   7.4 FFS analysis of corroded regions
   7.5 API RP 579 FFS evaluations
   7.6 Required thickness determination
   7.7 Evaluation of equipment with minimal documentation
   7.8 Reports and records

8. REPAIRS, ALTERATIONS AND RERATING OF PRESSURE VESSELS
   8.1 Repairs and alterations
   8.2 Rerating

9. ALTERNATIVE RULES FOR EXPLORATION/PRODUCTION VESSELS
   Not in the API 510 exam syllabus

APPENDICES
- APPENDIX A ASME CODE EXEMPTIONS
- APPENDIX B INSPECTOR CERTIFICATION
- APPENDIX C SAMPLE PRESSURE VESSEL INSPECTION RECORD
- APPENDIX D SAMPLE ALTERATION/RERATING FORM
- APPENDIX E TECHNICAL ENQUIRIES

Figure 2.1   API 510 contents/index

- Machinery, i.e. pumps, compressors, etc.
- Pipes and fittings

There are also some specific exemptions on size. Read the list in appendix A and relate them to Figs 2.2 and 2.3 below.

Appendix A (b6) gives an overall pressure temperature

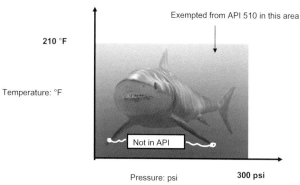

Figure 2.2   API 510 exemption: water under pressure

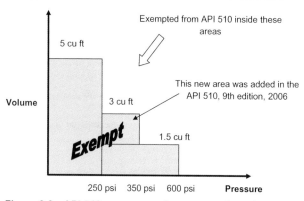

Figure 2.3   API 510 pressure–volume exemptions (appendix A (d))

exemption for vessels that contain water (or water with air provided as a 'cushion' only, i.e. accumulators).

Appendix A (b7) covers hot water storage tanks.

Appendix A (b8) gives a more general exemption based on minimum pressures and diameters.

Finally: **Appendix A (d)** covers a further general exemption based on pressure and volume.

Remember, section 1.2.2 at the front of API 510 only gives you half the story about exemptions. You have to look at the detail given in API 510 appendix A to get a fuller picture.

2.3 Section 3: definitions

Section 3.2: alteration

An *alteration* is defined as a change that takes a pressure vessel or component outside of its documented design criteria envelope. What this really means is moving it outside the design parameters of its design code (ASME VIII).

Note also how adding some types of nozzle connections may *not* be classed as an alteration. It depends on the size and whether it has nozzle reinforcement (in practice, you would need to check this in ASME VIII).

Section 3.6: authorized inspection agency

This can be a bit confusing. The four definitions (a to d) shown in API 510 relate to the situation in the USA, where the authorized inspection agency has some kind of legal jurisdiction, although the situation varies between states. Note this term *jurisdiction* used throughout API codes and remember that it was written with the various states of the USA in mind.

The UK situation is completely different, as the Pressure Systems Safety Regulations (PSSRs) are the statutory requirement. The nearest match to the 'authorized inspection agency' in the UK is probably 'The Competent Person' (organization) as defined in the PSSRs. This can be an independent inspection body or the plant owner/user themselves.

For API 510 exam purposes, assume that 'The Competent person' (organization) is taking the role of the authorized inspection agency mentioned in API 510 section 3.6.

Section 3.7: authorized pressure vessel inspector
This refers to the USA situation where, in many states, vessel inspectors *have to be* certified to API 510. There is no such legal requirement in the UK. Assume, however, that the authorized vessel inspector is someone who has passed the API 510 certification exam and can therefore perform competently the vessel inspection duties covered by API 510.

Section 3.9: condition monitoring locations (CMLs)
These are simply locations on a vessel where parameters such as wall thickness are measured. They used to be called thickness measurement locations (TMLs) but have now been renamed CMLs. CMLs pop up like spring flowers in a few places in API 510 and 572, with emphasis being placed on how many you need and where they should be.

Section 3.19: engineer
In previous editions of API 510, reference was made to the 'pressure vessel engineer' as someone to be consulted by the API inspector for detailed advice on vessel design. This person has now been renamed 'The Engineer'. There's progress for you.

Section 3.20: examiner
Don't confuse this as anything to do with the examiner that oversees the API certification exams. This is the API terminology for the *NDT technician* who provides the NDT results for evaluation by the API-qualified pressure vessel inspector. API recognizes the NDT technician as a separate entity from the API authorized pressure vessel inspector.

API codes (in fact most American-based codes) refer to NDT (the European term) as *NDE* (non-destructive examination), so expect to see this used throughout the API 510 training programme and examination.

An Introduction to API 510

Section 3.37: MAWP
US pressure equipment codes mainly refer to MAWP (maximum allowable working pressure). It is, effectively, the maximum pressure that a component is designed for. European codes are more likely to call it *design pressure*. For the purpose of the API exam, they mean almost the same, so you can consider the terms interchangeable.

Note how API 510 section 3.37 defines two key things about MAWP:

- It is the maximum gauge pressure permitted at the *top of a vessel* as it is installed (for a designated temperature). This means that at the bottom of a vessel the pressure will be slightly higher owing to the self-weight of the fluid (hydrostatic head). The difference is normally pretty small, but it makes for a good exam question.
- MAWP is based on calculations using the minimum thickness, *excluding the amount of the actual thickness designated as corrosion allowance*.

A significant amount of the exam content (closed-book and open-book questions) involves either the calculation of MAWP for vessels with a given amount of corrosion or the calculation of the minimum allowable corroded thickness for a given MAWP.

Section 3.53: repair
This is a revised definition added in the latest edition of API 510. It is mainly concerned with making a corroded vessel suitable for a specified design condition. If an activity does not qualify as an alteration then, by default, it is classed as a repair.

Section 3.54: repair organizations
API 510 has specific ideas on who is allowed to carry out repairs to pressure vessels. Look how definition 3.54 specifies four possible types of organization, starting with an organization that holds an ASME 'code stamp' (certificate

of authorization). This links in with the general philosophy of ASME VIII, requiring formal certification of companies who want to manufacture/repair ASME-stamped vessels.

Section 3.56: re-rating
The word *re-rating* appears frequently throughout API codes. Re-rating of the MAWP or MDMT (minimum design metal temperature) of pressure vessels is perfectly allowable under the requirements of API 510, as long as code compliance is maintained. In the USA, the API authorized inspector is responsible for re-rating a pressure vessel, once happy with the results of thickness checks, change of process conditions, etc. In the European way of working, this is unlikely to be carried out by a single person (although, in theory, the API 510 qualification should qualify a vessel inspector to do it). Re-rating may be needed owing to any combination of four main reasons – we will look at this in detail in Chapter 5.

Section 3.62: transition temperature
API codes are showing increasing acceptance of the problem of brittle fracture of pressure equipment materials. The new API 510 9th edition introduces the well-established idea of transition temperature, the temperature at which a material changes from predominantly ductile to predominantly brittle. As a principle, it is not advisable to use a material at an MDMT below this transition temperature (although we will see that there are possible 'get-outs' in the ASME VIII part of the syllabus) .

2.4 Section 4: owner/user/inspection organizations

Figure 2.4 summarizes the situation as seen by API.

Sections 4.1–4.2: responsibilities of user/owners
These sections are quite wide-ranging in placing an eye-watering raft of organizational requirements on the user/owner of a pressure vessel. This fits in well with the situation

An Introduction to API 510

in other countries where the owner/user ends up being the predominant duty holder under the partially sighted eye of the law.

The idea is that the owner/user should have a maintained QA/inspection/repair management system covering ... just about everything. There is nothing particularly new about the list of requirements of this (listed as section 4.2.1 *a* to *s*); they are much the same as would be included in an ISO 9000 audit or similar act of organizational theatre. They are also the same as those given in the API 570 Piping Inspection code. Note a couple of interesting ones, however.

Section 4.2.1(j): ensuring that all jurisdictional requirements for vessel inspection, repairs, alteration and re-rating are continuously met
Remember that the term *jurisdiction* relates to the legal requirements in different states of the USA. In the UK this would mean statutory regulations such as the PSSRs, HASAWA, COMAH, PUWER and suchlike.

Section 4.2.1(n): controls necessary so that only materials conforming to the applicable section of the ASME code are utilized for repairs and alterations
This is clear. It effectively says that only *code-compliant* material and procedures must be used for repairs and alterations if you want to comply with API 510. Note that (along with definition 3.3), it *does not specify exclusively* the ASME code; this is a significant change from previous API 510 editions which recognized only ASME as the 'applicable code'. You can think of this as a way of trying to make API 510 more relevant to countries outside the US, but remember that API 510 does not actually say this. The exam paper will be about what is written in the code, not your view of how it fits into the inspection world in other countries.

Reminder: API 510 says that: *only materials conforming to the applicable codes and specifications should be used for repairs and alterations.*

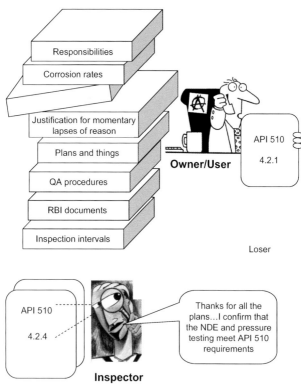

Figure 2.4 The balance of power

Section 4.2.1(0): controls necessary so that only qualified non-destructive examination (NDE) personnel and procedures are utilized

This means that API 510 requires NDE technicians to be qualified, although it seems to stop short of actually excluding non-US NDE qualifications. Look at section 3.27 and see what you think.

Some plant owner/users who have not read API 510 (why

An Introduction to API 510

should they, as they leave that to the inspector?) may need convincing that they are ultimately responsible for the long list of responsibilities in 4.2.1. However, they find out pretty quickly after a pressure-related incident.

Section 4.2.4: responsibilities of the API authorized pressure vessel inspector

This section appears in many of the API codes. The overiding principle (see Fig. 2.5) is that the API-certified pressure vessel inspector is responsible to the owner/user for confirming that the requirements of API 510 have been met. You will see this

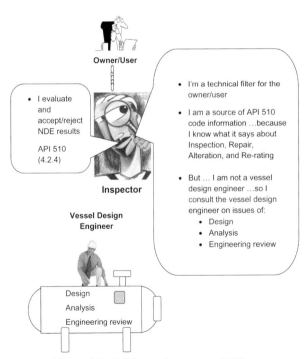

Figure 2.5 API inspector responsibilities

as a recurring theme throughout this book (and there will almost certainly be examination questions on it).

Section 4.2.4 places the requirements for candidates to have minimum qualifications and experience, before they are allowed to sit the API 510 exams (see appendix B where these requirements are listed).

Now, using your code, try to answer these familiarization questions.

2.5 API 510 sections 1–4 familiarization questions

Q1. API 510 section 1.2.2 and appendix A: exclusions
Which of these vessels is *excluded* from the requirements of API 510?

(a) A fired tubular heat exchanger in an oil refinery ☐
(b) A 200-litre air receiver at less than 0.5 bar gauge pressure ☐
(c) A separator vessel on an offshore platform (in USA waters) ☐
(d) All vessels operating at less than 250 psi ☐

Q2. API 510 section 1.2.2 and appendix A: exclusions
Which of these vessels containing steam is *excluded* from the requirements of API 510?

(a) A 100-litre vessel at 18 bar ☐
(b) A 100-litre vessel at 250 psi ☐
(c) A vessel of any size as long as the pressure is less than 300 psi ☐
(d) A vessel of any pressure as long as the capacity is less than 120 gallons ☐

Q3. API 510 section 2: references
Which API code (not in the API 510 syllabus) covers welding on equipment in service?

(a) API 579 ☐
(b) API 574 ☐
(c) API 2201 ☐
(d) SNT-TC-1A ☐

Q4. API 510 section 3.2: alterations

Which of these is *likely* to be classed as an *alteration* to a pressure vessel?

(a) Replacing the entire vessel head with one of the same design ☐
(b) Replacing existing nozzles with smaller ones ☐
(c) Replacing existing nozzles with larger ones ☐
(d) An increase in design pressure ☐

Q5. API 510 section 3.37: definitions of MAWP

In simple terms, MAWP means much the same as?

(a) 90 % design pressure ☐
(b) 150 % design pressure ☐
(c) Design pressure ☐
(d) Hydraulic test pressure ☐

Q6. API 510 section 3.37: definitions of MAWP

At what position is MAWP calculated for in a vertical pressure vessel?

(a) At the bottom of the vessel ☐
(b) At the top of the vessel ☐
(c) Halfway up the vessel ☐
(d) At the pressure gauge position, wherever it happens to be ☐

Q7. API 510 section 3.37: definitions of MAWP

A pitted vessel is measured at an average of 10 mm thick and has a 'designed' corrosion allowance of 1.6 mm. It is now installed in a corrosion-free environment. What thickness is used when calculating the MAWP?

(a) 13.2 mm ☐
(b) 11.6 mm ☐
(c) 10 mm ☐
(d) 8.4 mm ☐

Q8. API 510 appendix B3: inspector recertification

How often must an API 510-certified vessel inspector be 'recertified' by API?

(a) Every year ☐
(b) Every 3 years ☐
(c) Every 5 years ☐

(d) It depends on how many vessels the inspector has inspected ☐

Q9. API 510 section 4.4: responsibilities
During the repair of a pressure vessel, who is the API-certified vessel inspector responsible to?

(a) The owner/user of the vessel ☐
(b) API ☐
(c) The repair contractor ☐
(d) All the parties, as the inspector is acting in an 'honest broker' capacity ☐

Q10. API 510 section 4.2: API code compliance
During the repair of a pressure vessel, who is *ultimately* responsible for compliance with the requirements of API 510?

(a) The API-certified inspector ☐
(b) The owner/user ☐
(c) The repair contractor ☐
(d) It absolutely depends on the contractual arrangements in force ☐

Chapter 3

API 510 Inspection Practices (Section 5)

3.1 Introduction to API 510 section 5: inspection practices

Section 5 of API 510 contains many of the important principles on which the syllabus (and examination) is based. It is not really a stand-alone chapter; it relies on additional information included in sections 6 and 7 to give the full picture of what API considers is involved in the inspection of pressure vessels. Figure 3.1 shows the situation. This section has changed emphasis significantly since the previous API 510 edition; its main emphasis is now the existence and use of an inspection plan (written scheme of examination) linked with the application of risk-based inspection (RBI) techniques to help decide inspection scope and frequency. It also includes information on pressure testing, to link in with the requirements of ASME VIII.

3.2 Inspection types and planning

Section 5.1: inspection plans
Have a quick look through this. It is mainly commonsense about what should go in a vessel inspection plan. There is nothing in here that should be new to engineers who have worked with written schemes of examination (WSEs). It is, however, a good area for closed-book exam questions.

Section 5.2: risk-based inspection
This is a heavily expanded section compared to previous editions of API 510. Mention RBI in the world of inspection and it seems you just can't go wrong. Notice the two fundamental points:

- Inspection scope and frequency can be decided by

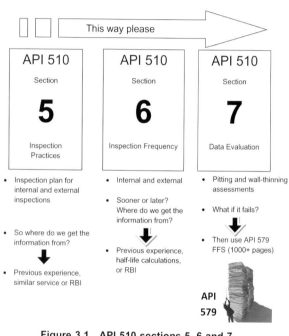

Figure 3.1 API 510 sections 5, 6 and 7

considering the *risk* that individual pressure vessels represent.
- Risk is determined by considering both probability of failure (POF) and consequences of failure (COF).

The content of this section 5.2 is taken from the document API RP 580: *Risk-Based Inspection*. This document is not in

API 510 Inspection Practices

the API 510 syllabus (it forms a supplementary examination and certificate in itself), but you are expected to know the summary of it that has been transplanted into section 5.2. Notice the breakdown:

- POF assessment
- COF assessment
- Documentation
- RBI assessment frequency

Section 5.3: preparatory work
Section 5.3 is mainly about good health and safety practice and commonsense. There is nothing in here that should be new to engineers who have worked on industrial sites. It is, however, a good area for an occasional closed-book exam question.

Section 5.4: Modes of deterioration and failure
This section is an introduction (only) to the types of failure and damage mechanisms (DMs) that can affect pressure vessels. As with so many of the API code clauses, it is a mixture of general descriptions and a few specifics. Note the general DM categories that are given in the list:

- General/local metal loss
- Surface-connected (breaking) cracking
- Subsurface cracking
- Microfissuring/microvoid formation
- Metallurgical changes
- Blistering
- Dimensional changes
- Material properties change

Most of these are covered in much more detail in API 571: *Deterioration Mechanisms* (this is part of the API 510 syllabus and we will be looking at it later in this book).

Section 5.5: general types of inspection and surveillance

This fairly general section introduces the different types of inspection that are commonly used for pressure vessels. In reality, there is very little information in here; most technical details come later in section 6. One clear requirement, however, is the need to assess the condition of linings and claddings, to guide the inspector's decision as to whether they need to be removed to inspect underneath.

Section 5.5.4: external inspection

This introduces the general requirements for external visual examination of pressure vessels. Note how it gives various different areas that should be assessed, including those for buried vessels. These are largely commonsense.

Section 5.5.6.1: CUI inspection

API codes like to warn against CUI (corrosion under insulation) and there are normally questions on it in the API 510 exam. They have recently revised the 'at-risk' temperatures for CUI to:

- Low carbon/alloy steels: 10 °F to 350 °F
- Austenitic stainless steels: 140 °F to 400 °F

Note that these are for systems that operate at a constant temperature. By inference, all systems that operate *intermittently* may be at risk from CUI whatever their temperature range.

Section 5.5.6.3: insulation removal

For vessels with external insulation, provided the insulation and cladding is intact and appears to be in good condition, API's view is that it is not necessary to remove the coating. It is, however, often good practice to remove a small section to assess the condition of the metal underneath.

Note the list of aspects to take into account when considering insulation removal:

- History

- Visual condition and age of the insulation
- Evidence of fluid leakage

This section introduces the principle that shell thicknesses in areas of CUI susceptibility (corrosion of the external surface) may be checked from the *inside* of a vessel during an internal inspection.

3.3 Condition monitoring locations (CMLs)

API 510 makes a huge fuss about CMLs (until recently referred to as thickness measurement locations (TMLs)). The change was made to recognize the fact that, in many systems, wall thinning alone is not the dominant damage mechanism. Other service-specific mechanisms such as stress chloride corrosion cracking (SCC) and high-temperature hydrogen attack (HTHA) are likely to be equally or more important.

Section 5.6 contains a page or so of commentary on good practice for selecting CMLs. Most NDE techniques are mentioned as being suitable, as long as their application is carefully chosen. This is followed by section 5.7.1, a well-defined list of NDE techniques and the type of defect they are best at finding. This is an important list for exam questions; the content also appears in other parts of the syllabus such as API 577 and ASME V.

Section 5.7.2: thickness measurement methods
Simple compression-probe ultrasonic testing (UT) is generally explained to be the most accurate method of obtaining thickness measurements. Profile radiography may be used as an alternative and in reality is often more useful. Note the requirements of section 5.7.2.3, which requires compensation for measurement inaccuracies when taking thickness measurements at temperatures above 65 °C (150 °F). This is covered in more detail in ASME V article 23.

3.4 Section 5.8: pressure testing

The requirement for doing a pressure test is often misunderstood, not least because of the fact that the mandatory requirement for it has been softened over the past 20 years or so. The situation in the current 9th edition of 510 is fairly clear, as follows:

- A pressure test *is* normally required after an *alteration*.
- The *inspector decides* if a pressure test is required after a *repair*.
- A pressure test is *not* normally required as part of a *routine inspection*.

API 510 gives no new requirements for test pressure; referring directly back to ASME VIII-I UG-98/99 requirements. Whereas in earlier editions (pre-1999 addendum) ASME VIII has used $1.5 \times$ MAWP as the standard multiplier for hydraulic test pressures, this was amended (1999 addendum and later) to the following:

- Test pressure (hydraulic) = $1.3 \times$ MAWP $\times$ ratio of material stress values
- Ratio of material stress values $= \dfrac{\text{Allowable stress at test temperature}}{\text{Allowable stress at design temperature}}$

Remember that this test pressure is measured at the highest point of the vessel. The allowable stress values are given in ASME II(d). Note that where a vessel is constructed of different materials that have different allowable stress values, the *lowest* ratio of stress values is used. You will see this used later in ASME VIII worked examples.

Section 5.8.6: test temperature and brittle fracture
US codes are showing an increasing awareness of the need to avoid brittle fracture when pressure testing of vessels. API 510 therefore now introduces the concept of *transition temperature*. To minimize the risk of brittle fracture, the test temperature should be at least 30 °F (17 °C) above the minimum design temperature (MDMT). There is no need to go above 120 °F (48 °C) as, above this, the risk of brittle fracture is minimal.

The temperature limitation is to avoid the safety risks that arise from brittle fracture of a vessel under pressure. Even when a hydrostatic (rather than pneumatic) test is performed, there is still sufficient stored energy to cause 'missile damage' if the material fails by brittle fracture.

Note also the requirement for temperature equalization; if the test temperature exceeds 120 °F, the test should be delayed until the test medium reaches the same temperature as the vessel itself (i.e. the temperatures have equalized out).

The hydrostatic test procedure

ASME VIII UG-99 (g) gives requirements for the test procedure itself. This is a fertile area for closed-book examination questions. An important safety point is the requirement to fit vents at all high points to remove any air pockets. This avoids turning a hydrostatic test into a pneumatic test, with its dangers of stored energy.

Another key safety point is that a visual inspection of the vessel under pressure is *not* carried out at the test pressure. It must be reduced back to MAWP (actually defined in UG-99 (g) as test pressure/1.3) before approaching the vessel for inspection. If it was a high-temperature test (> 120 °F, 48 °C), the temperature must also be allowed to reduce to this, before approaching the vessel.

Once the pressure has been reduced, all joints and connections should be visually inspected. Note how this may be waived provided:

- A leak test is carried out using a suitable gas.
- Agreement is reached between the inspector and manufacturer to carry out some other form of leak test.
- Welds that cannot be visually inspected on completion of the vessel were given visual examination prior to assembly (may be the case with some kinds of internal welds).
- The contents of the vessel are not lethal.

In practice, use of these 'inspection waiver points' is not very common. Most vessels are tested and visually inspected fully, as per the first sentences of UG-99 (g).

Quick Guide to API 510

A footnote to UG-99 (h) suggests that a PRV set to 133 % test pressure is used to limit any unintentional overpressure due to temperature increases. Surprisingly, no PRV set to test pressure is required by the ASME code. You just have to be careful not to exceed the calculated test pressure during the test.

Now try these familiarization questions.

3.5 API 510 section 5 familiarization questions

Q1. API 510 section 5.10.3: inspection of in-service vessels
Which of these in-service weld defects can be assessed by the inspector alone?

(a) Environmental cracking ☐
(b) Preferential weld corrosion in the HAZ ☐
(c) SCC ☐
(d) Straight, dagger-shaped, crack-like flaws ☐

Q2. API 510 section 5.8.5: pressure testing
When would a pneumatic test be used instead of a hydrostatic test?

(a) Because a vessel contains refractory lining ☐
(b) Because a vessel contains rubber or glass-reinforced plastic (GRP) linings ☐
(c) Because a vessel is constructed of 300 series stainless steel ☐
(d) Because a vessel is of riveted construction ☐

Q3. API 510 section 5.8.2: test pressure
What is the minimum code hydrostatic test pressure in the ASME VIII Div 1 1999 Addendum edition?

(a) MAWP, corrected for temperature ☐
(b) 130 % MAWP, corrected for temperature ☐
(c) 150 % MAWP, corrected for temperature ☐
(d) 110 % MAWP, excluding any temperature correction ☐

Q4. API 510 section 5.8.1.1: pressure testing
When is a pressure test normally required, without being specifically requested by an API inspector?

(a) During a routine inspection ☐

API 510 Inspection Practices

(b) Following a failure ☐
(c) After an alteration ☐
(d) After a repair ☐

Q5. API 510 section 5.7.1 (h): examination techniques
Acoustic emission techniques are used to detect:

(a) Leakage ☐
(b) Structurally significant defects ☐
(c) Stress and/or distortion ☐
(d) Surface imperfection in non-ferromagnetic material ☐

Q6. API 510 section 5.7.1 (a): examination techniques
Cracks and other elongated discontinuities can be found by:

(a) RT ☐
(b) VT ☐
(c) MT ☐
(d) PT ☐

Q7. API 510 section 5.6.3.1: CML selection
CMLs should be distributed:

(a) Appropriately over a vessel ☐
(b) In highly stressed areas ☐
(c) In areas of proven corrosion ☐
(d) Near areas of past failure ☐

Q8. API 510 section 5.5.6.3: CUI insulation removal
An externally lagged vessel has evidence of fluid leakage. Which of these is a viable option for an inspector who cannot insist that external lagging is removed?

(a) External UT ☐
(b) Internal UT ☐
(c) Thermography ☐
(d) None of the above; some insulation must be removed before the vessel can be approved for further service ☐

Q9. API 510 section 5.5.6.1: CUI susceptible temperature range
What is the CUI-susceptible temperature range of low alloy steel (e.g. $1\frac{1}{4}$ % Cr) vessels operating at constant (non-fluctuating) temperature

(a) -4 °C to 120 °C ☐
(b) -12 °C to 175 °C ☐

(c) 0 °C to 160 °C
(d) 6 °C to 205 °C

Q10. API 510 section 5.5.4.1.2: external inspection
External inspections are conducted to check for:

(a) External condition, hot spots and alignment issues
(b) Condition of the pressure boundary components only
(c) Health and safety hazards
(d) Wall thickness

Q11. API 510 section 5.5.3: on-stream inspection
All on-stream examinations should be conducted by:

(a) An inspector
(b) An examiner
(c) An inspector or examiner
(d) Someone familiar with the process conditions

Q12. API 510 section 5.2.1: probability assessment
A probability assessment should be in accordance with:

(a) ASME VIII or applicable codes
(b) API 580 section 9
(c) API 579 section 8
(d) API 581

Chapter 4

API 510 Frequency and Data Evaluation (Sections 6 and 7)

4.1 Introduction

Sections 6 and 7 of API 510 cover the subjects of *inspection interval frequency* (section 6) and *inspection data evaluation* (section 7). The contents of these sections are closely linked with, not only each other but also the previous section 5 covering inspection practices. Note the following points about these sections:

- They contain the main core material of API 510, particularly from the viewpoint of exam question content.
- All three of the sections have been added to, rearranged and changed in emphasis with each new edition of the code. They therefore have the characteristics of something that has grown organically rather than been designed recently from scratch. There is a focus and a logical order, of sorts, but this is surrounded by a mass of additional information contained in rather dense paragraphs of text.
- The title *data evaluation* predominates in section 7. Don't be misled by this – vessel inspection doesn't generally result in large data sets that need sorting and analysing – just think of it as describing what you do with inspection findings. Figure 4.4, later, shows the situation.

API 510 section 6 is a section of *principle*; it contains some of the major technical points of API 510 that appear in both open-book (calculation) and closed-book parts of the API 510 examination. These carefully presented eggs, golden or otherwise, contain much of the API view of the world on how and when vessels should be inspected. Section 6 is shorter than it was in previous editions, owing to the relocation of information about corrosion rates and pressure testing to other sections. The essential content can be distilled down to a handful of fairly straightforward background principles. Here they are:

Principle: Although section 6 provides guidelines; the API-certified vessel inspector retains a large amount of discretion as to what types of inspection are actually done on pressure vessels.

Principle: The general philosophy of section 6 is that the following inspection periods are used as a 'default level' for vessels, unless there is good reason to do otherwise:

- External visual inspection: 5 years
- Internal inspection or 'thorough' on-stream inspection: 10 years or half the remaining corrosion life (with a few exceptions)

Paradoxically, a lot of section 6 is then devoted to providing credible reasons for doing otherwise. The result is that section 6 is, in reality, promoting a risk-based inspection (RBI) approach to vessel inspection, but with a few 'boundaries' that should not be exceeded. If you read the section with this in mind, it will seem less muddled to you.

4.2 The contents of section 6

Figure 4.1 shows the contents list of section 6. It starts off with general information on the principles of inspecting vessels before use and after service changes and then moves on to the opportunities offered by RBI. It continually refers to the possibility of replacing internal inspections of vessels with on-stream inspections. Section 6.5 contains most of the key information.

Section 6.6, *PRVs*, contains information of a very general nature only and there is little in it that does not appear in the main PRV code (API 576). We will look at this later.

API 510 Frequency and Data Evaluation

Figure 4.1 What's in API 510 section 6?

Section 6.2: inspection during installation and service changes
Section 6.2 is a recent addition to the content of API 510. It makes the commonsense point that a pre-use inspection is required in order to collect base-line information that will be useful in future inspections. Look at 6.2.1.2 though, which makes clear that this does not have to include an internal vessel inspection; it is more about verifying what are essentially design and construction requirements. These are:

- Nameplate information
- Correct installation
- PRV settings

(See 6.2.1.1 list items (a) to (c).)

Section 6.3: RBI

This section contains little more than general **RBI** knowledge but opens the door to the key principle that the results of RBI evaluations can be used to override various API 510 requirements. Note the following two important points:

- The results of an **RBI** study can be used to change both the 5-year external and 10-year internal/on-stream 'default' inspection periods specified by sections 6.4 and 6.5 respectively.
- If you do the above and exceed the 10-year internal/on-stream inspection limit (stated in section 6.5), the **RBI** assessment must be revalidated at 10-year intervals (at least).

Figure 4.2 shows the basic principles of **RBI**. Don't get too excited about **RBI** in the context of API 510 – it has its own code (API RP 580), which is the subject of a separate ICP supplementary exam.

Section 6.4: external inspection

Prior to considering **RBI**, section 6.4 introduces the default level of 5 calendar years for external visual examination of pressure vessels. Note how it is expressed... *unless justified by an RBI assessment, for above ground vessels, an external inspection should be carried out at intervals not exceeding the lesser of 5 years or the required internal/on-stream inspection interval.* This mouthful relates to calendar years, and so applies whether the vessel is in continuous service or not (see section 6.4.2).

API 510 Frequency and Data Evaluation

THIS IS THE **OUTPUT** OF THE RBI ANALYSIS

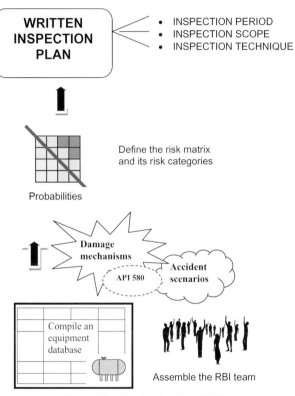

Figure 4.2 The basic idea of RBI

Section 6.5: internal and on-stream inspection (see Fig. 4.3)
Remember the principle in force here; *internal inspections* may be replaced by *on-stream* inspections. Section 6.5 contains the necessary qualifications to allow you to do

this. First, however, it provides the 'baseline' requirement for periodicity as follows.

The inspection interval should be not be more than:

- 10 years or
- Half the estimated remaining life

*Remember that on-stream inspections can be substituted for internal inspections …at the discretion of the inspector, and if you meet the list of conditions in 6.5.2.1

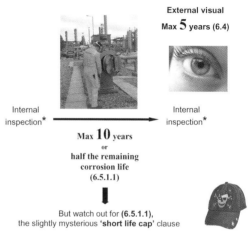

If the remaining life is <4 years, you can use the *full life* as the inspection period, rather than the half-life (up to a maximum of 2 years) … that's what it says …

SIMILAR SERVICE (6.5.2.3)

Figure 4.3   Internal/on-stream inspection periods

API 510 Frequency and Data Evaluation

If the calculated remaining life is less than 4 years, the inspection interval may be 100% of the remaining safe operating life, up to a maximum of 2 years. Let's call this the low-life cap, and not spend too much time trying to understand it.

For vessels not in continuous service that are blanked and purged so that no corrosion takes place, the inspection interval can be taken as a 10-year service life, but you have to be sure that the vessel is:

- Isolated from the process fluid and
- Not exposed to some other corrosive internal environment

Vessels that are not in continuous service and are not blanked and purged should be treated as normal continuous service vessels as above. Watch out for this as an exam question.

Although API 510 makes it quite clear that internal inspection is the preferred method of examination (particularly if there is localized corrosion or other types of damage) it then provides a healthy list of eight get-outs whereby it can be replaced with an on-stream inspection. See Fig. 4.3.

Section 6.6: pressure-relieving devices (PRVs)
Note the key content of section 6.6 covering the requirements of a PRV repair organization. They have to:

- Be experienced in valve maintenance
- Have a QC system and training programme that is fully documented
- Use qualified personnel
- Follow the requirements of API 576 when doing inspection and testing of PRVs

These points are commonsense. Most of the technical detail about PRVs comes later in API 576.

PRV inspection periods

Note the requirement for PRV inspection periods in section 6.6.2.2 of your code. The general 'default periods' are:

- 5 years or 10 years (depending on whether the process conditions are corrosive), and remember that these can (and should) be changed to fit in with the results of previous inspections and the dreaded RBI studies.

Now try these familiarization questions.

4.3 API 510 section 6 familiarization questions

Q1. API 510 section 6.2.1.1: inspection during installation

Which of these would not normally be included in a pre-use (installation) external inspection of a pressure vessel?

(a) Review of detailed design calculation ☐
(b) Check ladders and platforms ☐
(c) Wall thickness checks ☐
(d) Verify the nameplate correlates with the manufacturer's data report (MDR) ☐

Q2. API 510 section 6.2.1.2: inspection during installation

An inspector discovering that a newly installed vessel has a missing manufacturer's data report should:

(a) Perform an internal inspection of the vessel ☐
(b) Prohibit the vessel from being used ☐
(c) Remove the nameplate ☐
(d) Inform the owner/user ☐

Q3. API 510 section 6.3: risk-based inspection

An RBI assessment may be used to establish inspection intervals for:

(a) Internal inspections only ☐
(b) Internal and external inspections only ☐
(c) Internal, on-stream and external inspection ☐
(d) Vessels that have undergone a service change ☐

API 510 Frequency and Data Evaluation

Q4. API 510 section 6.4.1: external inspection period

How often should a vessel external inspection be performed on an above-ground vessel?

(a) 5 years ☐
(b) 10 years ☐
(c) Halfway through the calculated remaining life ☐
(d) It depends on the process ☐

Q5. API 510 section 6.5.1.1: internal inspection interval

What is the interval between internal inspections for a vessel with a projected remaining life of 30 years that is in continuous use?

(a) 5 years ☐
(b) 10 years ☐
(c) 15 years ☐
(d) At the discretion of the engineer and inspector ☐

Q6. API 510 section 6.5.1.1/6.5.1.2: internal inspection interval

A vessel has a projected remaining life of 15 years under its current regime of being in use 50 % of the time. The remainder of the time it is isolated from the process fluid and damage mechanisms by being filled with nitrogen. What is the internal inspection interval?

(a) 5 years ☐
(b) $7\frac{1}{2}$ years ☐
(c) 10 years ☐
(d) $12\frac{1}{2}$ years ☐

Q7. API 510 section 6.6: PRVs

Pressure-relieving devices should be inspected, tested and maintained in accordance with:

(a) ASME VIII or the 'applicable code' ☐
(b) API 576 ☐
(c) API 572 ☐
(d) API 520 ☐

Q8. API 510 section 6.6.2.2: PRV inspection intervals

The maximum test/inspection interval for PRVs in a 'typical process service' is:

(a) 2 years ☐

(b) 3 years ☐
(c) 5 years ☐
(d) 10 years ☐

4.4 Section 7: inspection data evaluation, analysis and recording

Whereas section 6 covers matters of principle, section 7 of API 510 is a cocktail of practical approximation and reasoned assumptions. It brings together most of the key concepts surrounding corrosion rate, remaining life, evaluation methods and fitness-for-purpose of corroded areas. It has been rearranged from previous editions of API 510, in which the above concepts were spread over several different sections, but the technical song remains much the same.

Note the breakdown of section 7 (you might find Fig 4.4 useful in visualizing what is in there).

Inspection data, evaluation, analysis and recording

7.1 Corrosion rate determination
7.2 Remaining life calculations
7.3 MAWP determination
7.4 Fitness-for-service analysis of corroded regions
7.5 API RP 579 FFS evaluations
7.6 Required thickness determination
7.7 Evaluation of existing equipment with minimal documentation
7.8 Reports and records

Think of sections 7.1 to 7.3 as fitting together into a set, suitable for dealing with uniform corrosion, with sections 7.4 and 7.5 acting as a general list of requirements to be used when assessing localized corrosion or defects in more detail.

Section 7.1: corrosion rate determination
API codes place stratospheric importance on the effects of wall thinning of pipes/vessels and the calculation of the maximum allowable working pressure (i.e. design pressure) that this horribly corroded item will stand. To this end, they use a mixed set of abbreviations and symbols to represent the

various material thicknesses at a condition monitoring location (CML). These look much more confusing than they actually are. Note the following definitions in API 510 section 7.1.1.1:

- $t_{initial}$ is the thickness measured at the first inspection (not necessarily when it was new) or the start of a new corrosion rate environment;
- t_{actual} is used to denote the actual thickness measured at the most recent inspection;

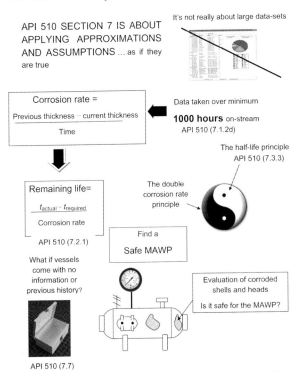

Figure 4.4a API 510 section 7: data evaluation; what's this all about?

THE HALF-LIFE / DOUBLE CORROSION RATE PRINCIPLE: API 510 (7.3.3)

THINK OF THESE AS JUST TWO WAYS OF EXPRESSING THE SAME THING

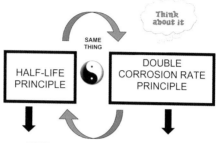

THEY ARE USED LIKE THIS...

When calculating remaining lifetime and the next inspection period:	When calculating safe MAWP 'now' (same as at the next inspection period)
Remaining life = $\dfrac{t_{actual} - t_{min}}{\text{Corrosion rate}}$	Using for example: $P = SEt/(R_i + 0.6t)$
Use the **single 'real'** corrosion rate	The t to use is t_{now} – **double the real corrosion loss expected**
Then **half** the lifetime to get the next inspection period	$t = t_{now} - (2 \times \text{'real' corrosion rate} \times \text{years to next inspection})$

Figure 4.4b API 510 section 7.3.3: the half-life/double corrosion rate principle

- $t_{previous}$ is the thickness measured at an inspection previous to another specified inspection;
- $t_{required}$ is a calculated value, rather than a measured one. It is the minimum (safe) required thickness in order to retain safely the pressure and (more importantly) meet the requirements of the design code (e.g. ASME VIII). This calculated required thickness excludes any specified corrosion allowance (it will be added on afterwards).

There is substantially less to these definitions than meets the eye. Just read them slowly and they will make sense.

API 510 likes to differentiate between the long-term (LT)

corrosion rate and the short-term (ST) corrosion rate. Again, this is nothing to get excited about. Figure 4.5 shows the idea in a simpler form.

A commonsense principle of API codes is that the most pessimistic corrosion rate (from those that are considered relevant) is used. It then falls to the API-certified pressure vessel inspector in conjunction with a corrosion specialist to decide which ones are relevant (that is what it says in section 7.1.1.1). For example, if there have been recent changes in

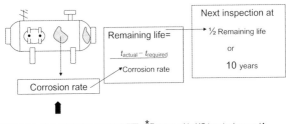

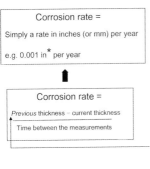

Figure 4.5 Corrosion rate definitions

process conditions, then the current short-term rate will be more relevant than the long-term rate experienced under the old process conditions.

Section 7.1.2: newly installed vessels: corrosion rate determination

For new(ish) vessels, it is obviously difficult to establish a valid corrosion rate. Section 7.1.2 gives four possible ways to 'estimate' it (see Fig. 4.6):

- Calculate it from data supplied about vessels in the same or similar service.
- Estimate it from the owner/user's experience.
- Estimate it from published data (e.g. NACE corrosion handbook).
- Measure it by taking on-stream thickness measurements after a minimum 1000 hours of operation (and keep it under review through time).

This section has little validity on its own; its main purpose is to support the subsequent section 7.3, where the objective is to go on to calculate MAWP in a corroded vessel or calculate a vessel's remaining lifetime and inspection period for a given MAWP.

Section 7.3: MAWP (maximum allowable working pressure) determination

We saw in Chapter 3 of this book how US pressure equipment codes mainly refer to MAWP (maximum allowable working pressure) as the maximum pressure that a component is designed for. European codes are more likely to call it *design pressure*.

Remember the two key things about MAWP:

- It is the maximum gauge pressure permitted at the top of a vessel as it is installed (for a designated temperature). This means that at the bottom of a vessel the pressure will be slightly higher owing to the self-weight of the fluid (hydrostatic head).
- MAWP is based on calculations using the minimum

API 510 Frequency and Data Evaluation

ESTIMATE THE CORROSION RATE FROM:

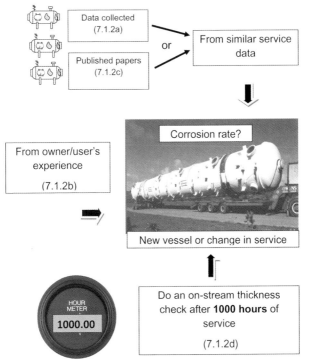

Figure 4.6 Corrosion rates for newly installed vessels

thickness, excluding the amount of the actual thickness designated as the corrosion allowance.

A significant amount of the exam content (closed-book and open-book questions) involves either the calculation of MAWP for vessels with a given amount of corrosion or the calculation of the minimum allowable corroded thickness for a given MAWP. Figure 4.7 shows the principle.

Finally, note the statement in section 7.3.1 about code editions. The idea is that MAWP calculations can be based on either the latest edition of the ASME code or the edition to which the vessel was built. This may be to fit in with the way that the system of compliance with the ASME code works in the USA, with a legal requirement for code compliance in most states.

Section 7.4: fitness-for-service analysis of corroded regions
This section is one of the core parts of API 510. Its content always appears in the examinations, in one form or another. Simplistically, it works on the view that corrosion may be either:

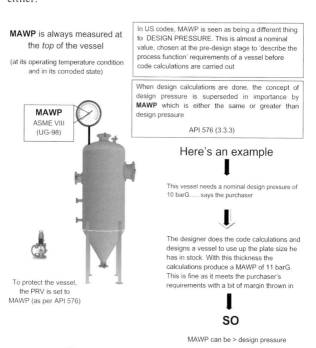

Figure 4.7   The principles of MAWP

API 510 Frequency and Data Evaluation

- uniform, so it may be difficult to see visually, or
- localized, i.e. some areas of material may be thinner than they appear, and it may be difficult to get true readings owing to surface roughness.

The following methods can be used to obtain the minimum thickness:

- UT: A-scan, B-scan and C-scan methods
- Profile RT
- Depth gauges

Section 7.4.2: evaluation of locally thinned areas
When considering the analysis of locally thinned areas, the concept of *average remaining thickness* is used. It is this average thickness that will be used to determine the corrosion rate, and subsequently the estimated life and frequency of inspection for the vessel. There are a couple of simple rules to follow.

Areas with no nozzles

For areas of significant corrosion, the remaining material thickness is calculated from equally spaced thickness measurements averaged over a length not greater than:

- For vessels up to 60-inch ID (inside diameter): half the vessel diameter or 20 in, whichever is less.
- For vessels greater than 60-inch ID: one third the vessel diameter or 40in, whichever is less.

Figure 4.8 shows the situation. Note the direction along which thickness readings are taken. If, as for most vessels, circumferential (hoop) stresses are the governing factor, then thickness readings will be averaged along the longitudinal direction (look at the statements in section 7.4.2.3).

Areas with nozzles

If the averaged area contains an opening (i.e. a nozzle) the situation is slightly different. The requirement is that the averaging area should not extend inside the limit of

Quick Guide to API 510

API 510 Section 7 concentrates on assessing corrosion in regions of circumferential (hoop) stresses, because that is the way that vessels under internal pressure tend to fail. The principle is that minimum thickness is assessed over an 'averaging longitudinal length'

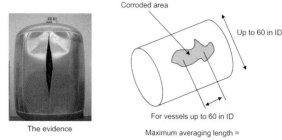

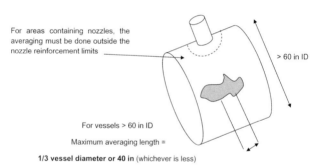

Figure 4.8   API 510 section 7.4.2.1: corrosion averaging

reinforcement defined in ASME VIII UG-35. We will look at this later when we consider ASME VIII.

Section 7.4.3: evaluation of pitting
API codes have well-defined ideas about the effect of pitting on vessel integrity. The clarity of the code sections covering this has varied in past code editions but the general principle is clear. Widely scattered pits are *not a threat to integrity* and may be ignored provided the following criteria are true:

- The remaining thickness left below any pit is more than half the $t_{required}$ thickness.

    AND, IF IT PASSES THAT

- The total *area* of the pits that are deep enough to eat into $t_{required}$ do not exceed 7 in² in any circle that is 8 in in diameter.

    AND, IF IT PASSES THAT...

- The sum of the *dimensions* of the pits that eat into $t_{required}$ along any straight line within an 8-in circle do not exceed 2 in.

You have to be careful of the API code definitions here as it uses the term 'corrosion allowance' to describe the amount of spare material left over and above $t_{required}$. Don't confuse this with the nominal corrosion allowance added to calculated thicknesses in ASME VIII. This is a totally separate idea. The thrust of the idea is that it is thinning below the $t_{required}$ thickness that threatens integrity. Figure 4.9 shows the principle.

This is not the only way that pitting/defects can be assessed. API also allows you to use the enthusiasm-sapping 1000+ pages of API 579. Note the specific comment in API 510 (section 7.4.4.1) that recommends API 579 for assessing ground areas where defects have been removed. Note also that the method of analysis given in ASME VIII division 2 appendix 4 may be used (see section 7.4.4.2). This is a complex method involving the determination of design stresses and should be carried out by an experienced pressure vessel engineer. The detail of this is outside the scope of the API 510 examination.

Corroded vessel heads

API 510 section 7.4.6 deals with the way to treat corrosion in vessel heads. It presents a couple of approximations to simplify the process, depending on the shape of the head. Figure 4.10 shows the situation.

API codes have well-defined ideas about the effect of pitting on vessel integrity. The basic principles are as follows:

> **You can ignore scattered pits if all the following criteria are true:**
>
> - The remaining thickness left below any pit is more than half the $t_{required}$ thickness (7.4.3a)
>
>   **And also**
>
> - The total area of pits **that are deeper than the 'corrosion allowance'(meaning that eat into $t_{required}$)** do not exceed 7 sq. in in any circle of 8 in diameter (7.4.3b)
>
>   **And also**
>
> - The sum of the dimensions of the pits **that are deeper than the 'corrosion allowance'(meaning those that eat into $t_{required}$)** along any straight line within an 8 in circle do not exceed 2 in (7.4.3c)

This is what API 510 (7.4.3) means

Watch the API definitions. The term *'corrosion allowance'* here means the amount of spare material left over and above $t_{required}$. Don't confuse this with any corrosion allowance mentioned in ASME VIII; this has a different context. The thrust of the idea is that it is thinning below the $t_{required}$ thickness that threatens integrity

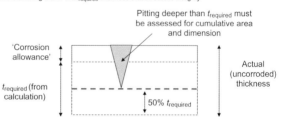

No pit must go so deep that it leaves a wall thickness of less than 50% $t_{required}$

Figure 4.9a Pitting interpretation (continues on next page)

Section 7.5: use of API 579 fitness-for-service evaluations
The main purpose of this section is to cross-reference the use of API 579. This is the API code covering fitness-for-service (or fitness-for-purpose). API 579 is a large document (1000 + pages) divided into many sections. It is used when vessels are in a damaged condition and have to be assessed to see if they are suitable for future use.

API 510 Frequency and Data Evaluation

If you are still struggling with API 510 clauses 7.4.3b and 7.4.3c), try this…

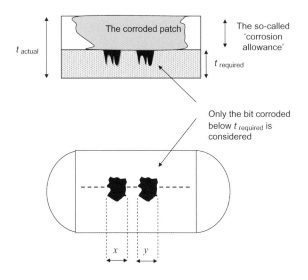

The sum of the dimensions ($x + y$) of areas corroded below $t_{required}$ along any straight line must not exceed 7 sq. inches area **and** 2 inches in 8 inches in the predominant stress direction (hoop stress direction normally governs)

Figure 4.9b   Pitting interpretation (continued)

You need to know of its existence (and what the sections cover) but the API 579 code itself is not in the API 510 examination syllabus. The main sections are (see Fig. 4.11):

- 7.5.3 Brittle fracture
- 7.5.4 General metal loss
- 7.5.5 Local metal loss
- 7.5.6 Pitting corrosion

- 7.5.7 Blisters and laminations
- 7.5.8 Weld misalignment and shell distortions
- 7.5. Crack-like flaws
- 7.5.11 Fire damage

There are three different levels of assessment for each of the sections in API 579, which become more complex as the level increases. The API inspection engineer may attempt level 1 but it is recommended that levels 2 and 3 would normally be carried out by experienced 'design' engineers. Other methods of defect assessment may be carried out, such as fracture mechanics evaluation of cracks, provided that you use an

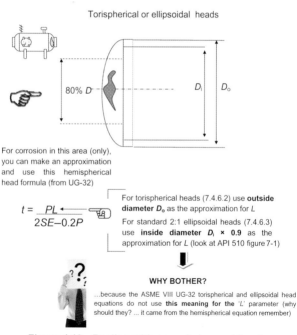

Torispherical or ellipsoidal heads

For corrosion in this area (only), you can make an approximation and use this hemispherical head formula (from UG-32)

$$t = \frac{PL}{2SE - 0.2P}$$

For torispherical heads (7.4.6.2) use **outside diameter D_o** as the approximation for L

For standard 2:1 ellipsoidal heads (7.4.6.3) use **inside diameter D_i × 0.9** as the approximation for L (look at API 510 figure 7-1)

WHY BOTHER?

...because the ASME VIII UG-32 torispherical and ellipsoidal head equations do not use **this meaning for the 'L' parameter** (why should they? ... it came from the hemispherical equation remember)

Figure 4.10 Dealing with corroded vessel heads

API 510 Frequency and Data Evaluation

Note the section numbers of API 579

These are common exam questions

(Not what's in them, just what section numbers they are)

API 579 section number.....

Figure 4.11   Some sections of API 579

established method and actually believe that these highly academic techniques have some relevance to the real world.

Section 7.7: evaluation of existing equipment with minimal documentation

This is a relatively new and expanding part of API 510. The objective is to give guidelines on what to do when dealing with vessels that have no nameplate details or construction date. Read through section 7.7 and note the main requirements as follows:

- You need to do retrospective design calculations to the appropriate edition of the ASME VIII code. Remember that major changes were made in 1999.
- Unidentified materials can be retrospectively qualified using UG-10(c) of ASME VIII or, as a default, use the material properties for the material SA-283 Grade C.
- If the RT grade (which defines the extent of radiography used during manufacture) is not known, use a joint efficiency of $E = 0.7$ for required thickness calculations.

After doing all this, it is the responsibility of the inspector to attach a new nameplate showing MAWP, maximum and minimum temperature (MDMT, minimum design metal temperature) and the date.

Now try these familiarization questions.

4.5 API 510 section 7 familiarization questions
Q1. API 510 section 7.1.1.1: corrosion rates
Short-term corrosion rates are typically determined by:

(a) The two most recent thickness readings ☐
(b) Any two sequential thickness readings ☐
(c) The two sequential readings showing the greatest corrosion ☐
(d) Any of the above, as the situation demands ☐

Q2. API 510 section 7.1.2: corrosion rate for vessels/changes in service
A vessel has just changed service and no published or 'similar service' data are available to predict the probable corrosion rate under the new service regime. What should the inspector do?

(a) Specify the first internal inspection after 1000 hours of operation ☐
(b) Specify thickness measurements after 1000 hours of operation ☐

(c) Assume a conservative corrosion rate of 0.5 mm per 1000 hours ☐
(d) Check the service conditions after 1000 hours to see if they are still the same; then specify an internal inspection ☐

Q3. API 510 section 7.4.2: evaluation of locally thinned areas

For vessels with an inside diameter of less than or equal to 60 inches, corroded wall thickness is averaged over a length not exceeding the lesser of:

(a) $\frac{1}{5}$ vessel diameter or 20 in ☐
(b) $\frac{1}{3}$ vessel diameter or 20 in ☐
(c) $\frac{1}{3}$ vessel diameter or 40 in ☐
(d) $\frac{1}{2}$ vessel diameter or 40 in ☐

Q4. API 510 section 7.4.2: evaluation of locally thinned areas

A vessel has an inside diameter of 90 inches. What is the maximum allowed averaging length for calculating corroded wall thickness?

(a) 20 in ☐
(b) 30 in ☐
(c) 40 in ☐
(d) 48 in ☐

Q5. API 510 section 7.4.2.3: evaluation of locally thinned areas

For an internally pressurized cylindrical vessel shell with no significant induced bending stresses (e.g. wind loads) corrosion is usually averaged along which plane?

(a) Circumferential ☐
(b) Axial (longitudinal) ☐
(c) Radial ☐
(d) Any plane containing the worst average corrosion ☐

Q6. API 510 section 7.4.4.2: advanced thinning analysis

As a general rule, the design stress used for an ASME VIII division 2 appendix 4 assessment (when used as an alternative to API 510) is:

(a) $\frac{2}{3}$ minimum specified yield strength ☐
(b) $\frac{1}{2}$ minimum specified yield strength ☐
(c) $\frac{1}{3}$ minimum specified yield strength ☐
(d) $\frac{2}{3}$ minimum specified ultimate tensile strength ☐

Q7. API 510 section 7.4.6: assessing corrosion in vessel heads

The effect of corrosion near the centre of vessel heads is calculated under API 510 using:

(a) The exact formulae used in ASME VIII-1 ☐
(b) The pitting assessment methodology set out in API 510 section 7.4.2 ☐
(c) Equations cross-referenced from API 579 ☐
(d) Approximation, to rules set out in API 510 section 7.4.6 ☐

Q8. API 510 section 7.4.6.2: corroded areas in vessel heads

For torispherical heads, the central portion can be assumed to be a hemisphere of radius equal to:

(a) The shell radius ☐
(b) The head knuckle radius ☐
(c) The head crown radius ☐
(d) The shell inner diameter ☐

Q9. API 510 section 7.4.6.3: corroded areas in vessel heads

For ellipsoidal heads the central portion can be assumed to be a hemisphere of radius equal to:

(a) The head inner radius x factor K_1 ☐
(b) The head crown radius x factor K_1 ☐
(c) Shell inside diameter x factor K_1 ☐
(d) Shell outside radius x factor K_1 ☐

API 510 Frequency and Data Evaluation

Q10. API 510 section 7.7: evaluation of vessels with minimal documents

For a pre-1999 pressure vessel that has drawings but no nameplate information the inspector and engineer should:

(a) Perform design calculations using a design factor of UTS/3.5 ☐
(b) Perform design calculations to the current edition of the ASME code ☐
(c) Perform design calculations based on the original construction code ☐
(d) Not perform any design calculations without the assistance of the original designer ☐

MEANWHILE, BACK IN THE PORTACABIN...
THEY'RE RUNNING TO THE WIRE

'Am I in the minority, or do I just not understand this "short-life" vessel inspection period cap?'

'Hmm, it sort of says ... when a vessel corrodes to within two years of its life, you can forget the half-life inspection interval and just use the full remaining life as the inspection period instead.'

'So we'd plan to inspect it exactly on the day when it is predicted to fail?'

'Let's see what it says in API 510 (6.5.1.1).'

'I'll have another look.'

'I've already looked ... we've got a vessel just like that outside, next to that wire fence.'

'Maybe there's a MAWP reduction built in somewhere, to reduce the risk as it gets near the day of its inevitable unspectacular demise.'

'Not necessarily. The life calculation is already based on a specific MAWP, in many cases the same as it always has been.'

'Maybe we can use the double corrosion rate thing?'

'I don't think it's possible; that's only used when calculating a MAWP and not remaining life according to API 510 (7.3.3).'

'Do you think we should follow it then?'

'Well, it's not mandatory to ignore the half-life inspection period for short-life vessels; it just says you can if you want to.'

'This is a big decision for me ... we'll ask the owner/users, it's their plant after all.'

'OK, and if the exam question asks what is the correct inspection period for a vessel that has only 2 years of remaining life, the answer is?'

'Two years API 510 (6.5.1.1).'

'... and if it has only one year of life remaining, the inspection period is?'

'One year.'

'Thank you, that's good enough for me.'

Chapter 5

API 510 Repair, Alteration, Re-rating (Section 8)

API codes are not only about inspection. Since their inception in the 1950s, the in-service inspection codes API 510, 570 and 653 have had the activities of repair, alteration and re-rating as a central part of their content. Worldwide, other in-service inspection codes do not necessarily follow this approach – re-rating, in particular, is uncommon in some countries. Many countries, as a result of legislation, experience or simple technical preference, do not do it at all. Remember, however, that the API 510 exam is strictly about what is in the code documents, not your experience or personal view, so you need to accept the API philosophy as set out in the codes.

5.1 Definitions

The three important definitions are those for repair, alteration and re-rating. Within the confines of the API codes these three definitions are almost a subject in themselves. They are not difficult – just a little confusing. The easiest way to understand them is to start with the definition of *alteration*.

Note two points:

- There are a specific set of circumstances that define the term *alteration*.
- If any grinding, cutting or welding activities are done that do not meet the requirements to be an alteration then they are defined as a *repair*.

The activities that define an alteration are best understood by looking at Fig. 5.1. Note how an alteration must involve a physical change with design implications that affect the pressure-retaining capability (i.e. the 'pressure envelope').

Quick Guide to API 510

AN ALTERATION IS:

A physical change that has *design implications* that weren't considered when it was originally manufactured

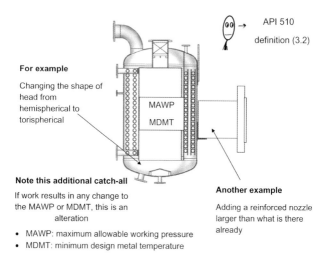

API 510 definition (3.2)

For example

Changing the shape of head from hemispherical to torispherical

MAWP
MDMT

Note this additional catch-all

If work results in any change to the MAWP or MDMT, this is an alteration

- MAWP: maximum allowable working pressure
- MDMT: minimum design metal temperature

Another example

Adding a reinforced nozzle larger than what is there already

> By default, work that is not considered an alteration is classed as a *repair*

Figure 5.1 Vessel alterations

This is set out in definition 3.2 of API 510 and is fairly easy to understand.

Slightly more difficult to accept is what is *not* defined as an alteration. Figure 5.2 shows the situation. Examination questions normally revolve around these definitions and may be open- or closed-book types.

5.2 Re-rating

This normally involves raising or lowering the design temperatures (MDMT) or design pressure (MAWP), or

API 510 Repair, Alteration, Re-rating

both. These are several different scenarios that may lead to a re-rating. Figure 5.3 shows the situation.

A re-rating may be necessary if a vessel is badly corroded. Serious thinning may require the MAWP to be reduced. Alternatively, if a process change involves a vessel requiring an increase or decrease (i.e. MDMT related) in temperature, then a re-rating will be required to ensure the vessel is safe under the new conditions.

An overriding principle is that any vessel subject to an

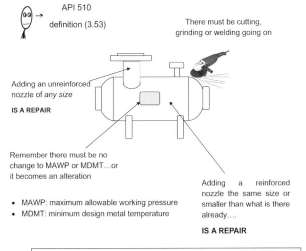

Figure 5.2 Vessel repairs

Quick Guide to API 510

THE FOUR REASONS FOR RE-RATING A VESSEL

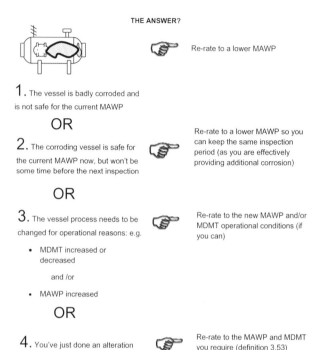

Figure 5.3 **Reasons for re-rating**

alteration has then to be re-rated. This makes sense as the alteration is, by definition, design related, so any changes in design condition has to be formalized by re-rating the vessel.

Watch out for exam questions about the *code edition* to which a re-rating is carried out. As a matter of principle, the primary objective is to re-rate a vessel to the *latest edition* of the code to which it was built (see Fig. 5.4). If this is possible, it will obviously incorporate any improvements in the code that have been introduced since the vessel was built. Codes do develop through time; allowable stress values, for example, may be increased or decreased based on new material developments or experience of failures.

As an alternative, if it is not possible to re-rate to the most up-to-date code edition, then the idea is to re-rate to the edition to which the vessel was built, as long as jurisdiction/ statutory requirements allow it.

You can see the two re-rating scenarios above set out in clause 8.2.1(b) of API 510. Note the mention of code cases

RE-RATING: WHICH CODE EDITION?

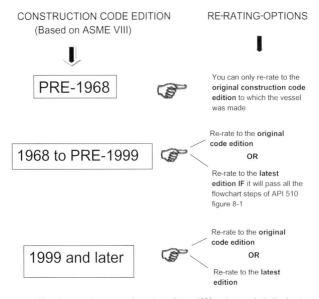

Note how re-rating a vessel constructed to a 1999 or later code is the least restrictive. This is probably because it already incorporates design changes made in the 1999 edition, bringing it more in line with current thinking on design formulae and material properties

Figure 5.4   Re-rating options

COMPARE THIS WITH:
API 510 FIGURE 8-1: RE-RATING

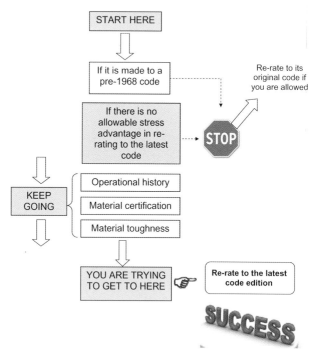

Figure 5.5 The re-rating flowchart

2290 and 2278 – these are the results of two code 'case study changes' relating to relevant editions. That is all you need to know for the purposes of the API 510 exams. The other relevant area to look at is Fig. 8.1 of API 510. This is simply a list of steps that must be complied with if you want to re-rate a vessel designed to ASME VIII code edition 1968 to pre-1999 to the latest edition of ASME VIII. Figure 5.5

API 510 Repair, Alteration, Re-rating

summarizes the situation – if you can understand this figure there is no need to worry too much about the wording of API 510 section 8.2.1(b).

Notice how the flowchart of Fig. 8.1 in API 510 works. Remember that the objective is to re-rate a vessel built to a 1968 to pre-1999 ASME VIII code using the revised material properties of the current code edition (e.g. 2007). The chronological steps start from the top centre of the flowchart with the objective being to progress vertically downwards, ending up at the rectangular box nearer the lower right of the page. This is the 'destination' box that allows you to legitimately re-rate the vessel to the latest code edition. Note how the vertical steps include things relating to mainly the material properties of strength and toughness (resistance to impact). This is to ensure that older or substandard materials cannot get through. Equally importantly, note the two boxes near the middle of the flowchart where it is necessary to confirm that the material has not been degraded (by corrosion, age-hardening, temper embrittlement, creep, etc.). This can sometimes be the most difficult step to fulfil, as the operational history of plant is so often incomplete. Don't worry about these practicalities for the API 510 exam – the exam questions are artificially simplified and often just replicate the exact wording in the flowchart 'boxes'.

5.2.1 What happens after a re-rating?

There are two main issues here: pressure testing and nameplates. Look at API 510 section 8.2.1(d), which mentions pressure testing – it confirms that pressure testing is 'normally' required, but not actually mandatory, as you are allowed to substitute it with 'special NDE' (section 8.2.1(d)(2)) as long as this is acceptable to the pressure vessel engineer. If you do perform a pressure test, remember the situation regarding hydraulic test pressure:

- ASME code edition before 1999: test pressure = $1.5 \times$ MAWP

- ASME code edition 1999 and later: test pressure = 1.3 × MAWP

The awkward scenario of what to do if you want to re-rate a vessel to a 1999 or later code, when it was built to an earlier (1968 to pre-1999) code with the higher MAWP multiplier, is covered in clause 8.2.1(d). Here the vessel has already been tested to higher than the '1999 and later' 1.3 × MAWP multiplier, so there is no need for an additional test.

Nameplates are fairly straightforward. API 510 section 8.2.2 explains that a re-rating requires an *additional* nameplate to be fitted (not a replacement). Alternatively, it is acceptable to add additional stamping. Either must be witnessed by the inspector to make the re-rating valid. This is a common examination question.

5.3 Repairs

Under the logic of API 510, any welding cutting or grinding operation that does not qualify for being an alteration (see Fig. 5.2) is, by default, classed as a repair. The main objective of API 510 with respect to repairs is to give technical guidelines as to how these repairs may be carried out rather than to specify any restrictive tests associated with them. Looking back to API 510 clause 5.8.1.1, you can see that a pressure test is not mandatory following a repair – it is at the discretion of the inspector. Surprisingly, it is not absolutely mandatory after an alteration either – clause 5.8.1.1 backs off slightly and says only that it is 'normally required'. That is not the same as it being mandatory. Expect exam questions to simply refer to the exact wording of these code clauses, rather than requiring you to interpret them or explain the logic behind them.

5.3.1 Repair techniques

Over recent code editions, API 510 has persistently increased its technical coverage of vessel repairs. This makes sense as 'repair' actually appears in the title of API 510. Most of the additions have been to section 8 – evidenced by the fact that

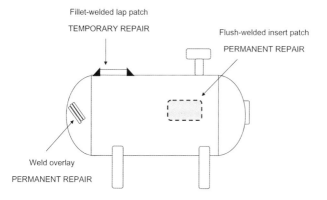

Control of weld repair activity:

- Avoidance of brittle fracture (material choice)
- Avoidance of weld cracking (material choice and heat input)

Figure 5.6 API 510 section 8: repair issues

it now contains up to seven levels of numbering hierarchy (e.g. clause 8.1.6.4.2.1.4), leaving it a little out of balance with other sections of the document. As a result of this growth, API 510 section 8 contains a lot of technically valid and useful points – perhaps several hundred of them. The problem is that they are contained in dense clauses of text with few explanatory pictures, so they don't exactly jump out of the page at you.

Figure 5.6 below is an attempt to summarize the situation. Look how the main issues are divided into four as follows:

- Temporary versus permanent repairs (which is which)
- Specific requirements for temporary repairs
- Specific requirements for permanent repairs
- Technical requirements (and restriction) of welding techniques – with the objective of avoiding two main generic problems – brittle fracture and excessive heat input.

Figure 5.7 shows some key technical points about weld overlay repairs while Fig. 5.8 shows the API 510 views on approvals and authorization of repairs.

Now try the familiarization questions in section 5.4.

REPAIR METHOD **APPROVED** BY THE **ENGINEER AND INSPECTOR**
8.1.5.4.1

REPAIR **MONITORED** BY **INSPECTOR**
8.1.5.4.3

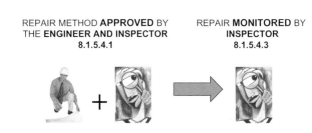

Previous exposure to sour service (H_2)? Leave for 24 hours after repair to allow H_2 outgassing causing delayed cracking. Do PT after cooling and UT from back side after 24 hours if base metal is P3, P4 or P5

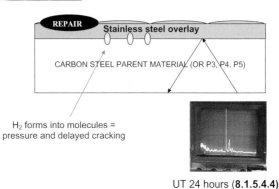

UT 24 hours (**8.1.5.4.4**)

Figure 5.7 API 510 section 8.1.5.4: weld overlay repairs

API 510 Repair, Alteration, Re-rating

APPROVAL: Meaning review of *what is to be done*
(8.1.2.1)

	APPROVAL OF METHOD (8.1.2.1)		
	Repairs	Alterations	Work in progress for either
Engineer	ONLY Fillet welded patches 8.1.5.1.2.1, Lap band repairs 8.1.5.1.3, Repair to SS (only) weld overlay 8.1.5.4.1	Yes	Not required
Inspector	Yes	Yes	Yes

AUTHORIZATION: Meaning *OK for work to proceed*

AUTHORIZATION TO PROCEED (8.1.1)		
	Repairs	Alterations
ASME VIII Div 1	Inspector	Inspector and Engineer
ASME VIII Div 2	Inspector and Engineer	Inspector and Engineer

Inspector may give *prior general* authorization for limited or routine repairs provided that:

- It is for a specific vessel
- The inspector is happy with the repair contractor's competence
- The repair type does not require pressure testing (e.g. weld overlay that does not require PWHT)

Figure 5.8   Who approves repairs and authorizations?

5.4 API 510 section 8 familiarization questions

Q1. API 510 section 8.1.1: repairs and alterations
What repair and alteration activities is the API inspector able to authorize alone (before the repair organization starts work) without reference to the engineer?

(a) Repairs to ASME VIII division 1 vessels ☐
(b) Repairs or alterations to ASME VIII division 1 vessels ☐
(c) Repairs to any ASME VIII vessels ☐
(d) Repairs or alterations to any ASME VIII vessels ☐

Q2. API 510 section 8.1.5.1.1: temporary repairs
Temporary repairs may remain in place past the first opportunity for replacement if:

(a) The engineer gives approval ☐
(b) The inspector gives approval ☐
(c) The details are properly evaluated and documented ☐
(d) All of the above ☐

Q3. API 510 section 8.1.5.1.2: temporary repairs: fillet-welded patches
Generally speaking, fillet-welded patches are unlikely to be suitable for providing a temporary repair to parts for a pressure vessel containing:

(a) Extensive corrosion ☐
(b) Deep corrosion ☐
(c) Cracks ☐
(d) All of the above ☐

Q4. API 510 section 8.1.5.1.3: temporary lap band repairs
The circumferential fillet welds attaching a lap band repair to a vessel shell should be designed with an assumed joint efficiency of:

(a) 1 ☐
(b) 0.7 ☐
(c) 0.45 ☐
(d) Appropriate to the applicable design code ☐

API 510 Repair, Alteration, Re-rating

Q5. API 510 section 8.1.5.2.2: flush insert plate repairs

Flush insert patches may be used to repair corroded shell plates as long as:

(a) Welds are full penetration ☐
(b) PT/MT is performed on the completed welds ☐
(c) PT/MT is performed on weld root runs and the completed welds ☐
(d) Welds have a minimum joint efficiency of 0.8 ☐

Q6. API 510 section 8.1.5.3.2: weld overlay repairs

A weld overlay repair stands 'proud' of the surface of the vessel shell. The inspector should:

(a) Accept the weld ☐
(b) Specify that the increased thickness be blended to a 3:1 taper ☐
(c) Seek approval from the engineer ☐
(d) Reject the weld ☐

Q7. API 510 section 8.1.5.4.4: repair to P3, P4 and P5 stainless steels

Weld repairs to P3, P4 and P5 stainless steels in any service require specific UT testing for:

(a) Immediate post-weld cracking in the HAZ ☐
(b) Outgassing problems ☐
(c) Delayed cracking in the base metal ☐
(d) Type IV creep cracking, delayed by 24 hours after welding ☐

Q8. API 510 section 8.1.6.4: PWHT

The API 510 approach to PWHT is that:

(a) The ASME code is mandatory ☐
(b) The applicable construction code is mandatory ☐
(c) Approved alternative procedures to those in construction codes are allowed ☐
(d) The engineer and inspector make the final decision on what is acceptable ☐

Q9. API 510 section 8.1.6.4.1(c): local PWHT

What is the minimum preheat that must be maintained during a repair that will have local PWHT?

(a) 150 °C ☐
(b) 150 °F ☐
(c) 300 °C ☐
(d) It depends on the material thickness ☐

Q10. API 510 section 8.1.6.4.2.3(b): preheat method in lieu of PWHT

CD welding in lieu of PWHT can only be used on:

(a) P1 and P3 steels ☐
(b) P1, P3 and P4 steels ☐
(c) P1 groups 1, 2, 3 and P3 groups 1, 2 steels ☐
(d) P1 groups 1, 2, 3 and P4 group 1 steels ☐

Chapter 6

API 572 Inspection of Pressure Vessels

6.1 API 572 introduction

This chapter is about learning to become familiar with the layout and contents of API 572: *Inspection of Pressure Vessels (Towers, Drums, Reactors, Heat Exchangers and Condensers)*. API 572 is a well-established document (it is still on its 2001 edition) with its roots in earlier documents published by the American Refining Industry (IRE). It is more a technical *guide document* rather than a code, as such, but it does perform a useful function in supporting the content of API 510.

Note the following five points about API 572:

Point 1. It has a *very wide scope* as evidenced by its title, which specifically mentions the types of vessels that it covers: *Towers, Drums, Reactors, Heat Exchangers and Condensers*. This wide scope is evident once you start to read the content; it refers to all these types of equipment and the materials, design features and corrosion mechanisms that go with them.

Point 2. API 572 introduces various *corrosion and degradation* mechanisms. As you would expect, these are heavily biased towards the refining industry, with continued emphasis on sulphur/H_2S-related corrosion mechanisms and cracking. In general, although it provides description and discussions on corrosion, API 572 acts only as *an introduction* to these corrosion mechanisms, leaving most of the detail to be covered in API 571.

Point 3. It is *downstream oil industry orientated* (not surprising as it is an API document). Its main reference is to the downstream oil sector. *Downstream* is a term commonly used to refer to the part of the industry involved in the selling and distribution of products derived from crude oil (gas, petrol, diesel, etc.). The types of equipment covered by the code can therefore include oil refineries, petrochemical plants, petroleum products distributors, retail outlets and natural gas distribution companies. These can involve thousands of products such as gasoline, diesel, jet fuel, heating oil, asphalt, lubricants, synthetic rubber, plastics, fertilizers, antifreeze, pesticides, pharmaceuticals, natural gas and propane, etc.

Note that the *upstream* oil sector (i.e. exploration and production (E&P) equipment) is not overtly covered in API 572. E&P vessels are specifically covered in API 510 section 9 but are excluded from the API 510 exam syllabus.

Point 4. This refers to quite a few *related codes* that are not in the API 510 exam syllabus (API 660/661 for heat exchangers, API RP 938/939/941, etc., and others; see API 572 section 2 on page 1 of the code). These provide technical details on specific subjects and problems but don't worry about them. You need to know that they exist but you do not need to study them for the API 510 examination.

And finally, the most important point. API 572 *is all text and technical descriptions*, accompanied by explanatory photographs of a fairly general nature. It contains no calculations. This means that most examination questions about API 572 in the API 510 certification exam will inevitably be *closed book*. The downside to this is that API 572 contains several thousands of separate technical facts, giving a large scope for the choice of exam questions.

All this means that you need to develop a working familiarity with the technical content of API 572, treating it as essential background knowledge for the API 510 syllabus, rather than as a separate 'stand-alone' code in itself. We will look at some of the more important areas as we work through the code.

6.2 API 572 section 4: types of pressure vessels

API 572 section 4 is little more than four pages of general engineering knowledge. It provides basic information about types of pressure vessels and their methods of construction. Note the points below.

Section 4.1: types of pressure vessels

A pressure vessel is defined as a container designed to withstand internal or external pressure and is designed to ASME VIII or other recognized code (section 4.1). Note the 'cut-off point' at 15 psi gauge pressure. This fits in with the 15 psi minimum pressure limit we saw previously in API 510 appendix A.

A vessel is most commonly a cylinder with heads of various shapes, such as:

- Flat
- Conical
- Toriconical
- Torispherical
- Semi-ellipsoidal
- Hemispherical

We will look at the calculations associated with some of these shapes later in this book.

Cylindrical vessels can be both vertical and horizontal and may be supported by:

- Columns (legs)
- Cylindrical skirts
- Plate lugs attached to the shell

Spherical vessels may be similarly supported by:

- Columns (legs)
- A skirt
- Resting on the ground (either partially or completely)

Section 4.2: methods of construction
Most vessels are of fully welded construction. There may be some old riveted vessels remaining but they are not very common.

The cylindrical shell courses (or *rings*) are made by rolling plate material and then welding the longitudinal joint. The courses are then assembled by circumferentially welding them together to give the required length of vessel. Hot forging of cylinders can be used, as this produces a seamless ring. Multilayer methods in which a cylinder is made up of a number of concentric rings can be used for heavy-wall vessels and items subject to high pressure. This is a very expensive method of fabrication.

Heads are made by forging or pressing, either hot or cold, from a single piece of material or built up of separate 'petal' plates.

6.3 API 572 Section 4.3: materials of construction

Section 4.3 summarizes the types of materials commonly used for pressure vessels. Treat this as general information only; materials are described in much more detail in ASME VIII and API 577 covered in other chapters of this book. The material categories are:

1. Carbon and low alloy steels
2. Stainless steels
   - Ferritic (13% Cr)
   - Austentic (18% Cr 8% Ni)
   - Duplex (25% Cr 5% Ni)
3. Non ferrous
   - Nickel alloys
   - Titanium
   - Aluminium
   - Copper
4. Lined vessels (a low-cost carbon steel base material with corrosion-resistant lining)

- Roll bonded
- Explosion bonded
- Welded sheets (API 572 Fig. 4)
- Weld overlay
5. Non-metallic liners (used for corrosion resistance or insulation)
    - Brick
    - Concrete
    - Rubber
    - Glass
    - Plastic

Section 4.4: internal equipment
Some vessels have no internal parts while others can have the following:

- Baffles
- Distribution trays
- Mesh grids
- Packed beds
- Internal support beams
- Cyclones
- Pipes
- Spray nozzles

Section 4.5: uses of pressure vessels
These are the uses indicated in API 572 section 4.5:

- To contain the process stream
- Reactors (thermal or catalytic)
- Fractionators (to separate gases or chemicals)
- Surge drums
- Chemical treatment vessels
- Separator vessels
- Regenerators

This is well short of being an exhaustive list, so treat it as 'general knowledge' information only.

6.4 API 572 sections 5, 6 and 7

Taken together, these comprise less than one full page of text. Read through them and note the following:

- The cross-references to the vessel construction code **ASME VIII** divisions 1 and 2 (make sure you understand what each division covers, although the API 510 syllabus is concerned with division 1 vessels only).
- The cross-references to **TEMA** (Tubular Exchangers Manufacturers Association) and **API 660/661**, the construction codes used for heat exchangers and condensers.
- The references to **OSHA** (Occupational Safety and Health Administration) and **NB-23** (The National Board vessel inspection code, in some states of the USA used as an alternative to API 510). You don't need to study these documents, just recognize that they exist.

Note the sentence hidden away in API 572 section 7.3 saying that periods of repair/replacement (and inspection) are based on corrosion rates and remaining corrosion allowances. This statement summarizes the entire philosophy of the API inspection codes and appears in various guises in all the codes included in the API 510 exam syllabus.

6.5 API 572 section 8: corrosion mechanisms

This is one of the most significant sections of API 572 but in reality there is very little in API 572 section 8 that is not covered in as much, or more, detail in API 571. Section 8 provides a few more photographs and adds explanations of a few additional degradation mechanisms (DMs) but there is little that is fundamentally new.

Three new DMs that *are* introduced are given in section 8.3.3: dealloying. They are:

- Dezincification (affects brass, which is a Cu/Zn alloy)
- Dealuminization (affects aluminium brasses or bronzes)
- Denickelification (affects cupronickel heat exchanger tubes or Monel metals)

API 572 Inspection of Pressure Vessels

These affect copper alloys and are all related to the leaching out of alloy constituents (the alloys are obvious as they appear in the DM title).

API 572 section 8.3.7 introduces another new DM: hydriding of titanium alloys. This is a type of embrittlement caused by absorption of hydrogen. Embrittlement is an important theme of API 571/572 and is evidence of the emphasis placed on refinery-type DMs.

Apart from these, all the other DMs in API 572 appear in API 571. They also appear in some other API codes (e.g. API 570: *Inspection of Pipework*).

The final 'new' part of Section 8 is 8.5: faulty fabrication. This is an unusual subject to cover in a chapter entitled 'Corrosion Mechanisms', but it only covers half a page and contains some useful points on vessel problems that have their root in the manufacturing stages. The causes are divided into (see section 8.5 on page 16 of API 572):

- Poor welding
- Incorrect heat treatment
- Wrong dimensions
- Incorrect installation
- Incorrect fit (assembly)
- Incorrect materials

Note the last one, *incorrect materials* (API 572 section 8.5.7). This short section mentions the advantages of positive material identification (PMI), carried out using a 'Metascope' or X-ray hand-held analyser. There is actually a dedicated API code for PMI techniques (API 578) but this is not part of the API 510 syllabus. It is, incidentally, part of the API 570 *in-service inspection of pipework* syllabus, presumably because someone has decided that the use of incorrect materials is more common in pipework components than in vessels.

6.6 API 572 section 9: frequency and time of inspection

This is a short section (less than two pages). It provides fairly general explanations of the principles of inspection frequency and reintroduces (once again) the half-life concept of API 510. Note the following two key points hidden away in the text:

- Section 9.2 contains the general statement of principle that 'on-stream inspections can be used to detect defects and measure wall thickness' (it is in 9.2 (d)). This is reinforcement of the general principle of API 510 that internal 'shutdown' examinations of vessels are *not absolutely essential* and may be replaced, where applicable, with a good-quality on-stream inspection using equipment able to detect defects and measure the wall thickness.
- Section 9.4: alternative rules for exploration and production (E&P) vessels is little more than an acknowledgement that E&P vessels may need a more risk-based approach. The text is much the same as that in API 510, and is not in the exam syllabus.

6.7 API 572 section 10: inspection methods and limitations

Here are some key points about API 572 section 10.

Section 10.2: safety precautions and preparatory work
While not particularly technically orientated, this short subsection is a fertile source of closed-book exam questions. Safety questions are always popular in examination papers so it is worth looking at this section specifically in terms of identifying content that could form the subject of an exam question. Note how continued emphasis is placed on safety aspects such as vessel isolations, draining, purging and gas testing.

API 572 Inspection of Pressure Vessels

Section 10.3: external inspections

This is a long section containing a lot of good-quality technical information on the external inspection of vessels. It contains 13 subsections that work through the physical features of vessels, providing guidance on what to inspect. The section complements section 6 of API 510, but goes into much more technical detail.

Remember that the emphasis of section 10.3 is on *external inspections*; don't confuse this with the more detailed *on-stream inspection* that, as a principle of API 510, can be used, where suitable, to replace an internal examination. In practice, this will involve advanced NDT techniques (corrosion mapping, eddy current, profile radiography, etc.) or similar. Strictly, this is not what section 10.3 is about; it restricts itself to more straightforward visual inspections.

Section 10.4: internal inspections

Section 10.4 goes into much more detail than previous sections of API 572. Spread over 7 to 8 pages, it provides a comprehensive technical commentary on the techniques for internal inspection of vessels. Once again, it is all qualitative information (there are no calculations) restricting the content to mainly *closed-book examination questions.*

From an API 510 examination viewpoint, the difficulty with section 10.4 is its wide scope. It covers subjects relating to general pressure vessels but intersperses these with techniques and corrosion mechanisms relating to specific refining industry applications (fractionating towers containing trays, low chromium alloy hydroprocessing units and similar).

Before we look at some simple questions, note the overall structure of section 10.4. It addresses things in the following order:

10.4.1 General
10.4.2 Surface preparation
10.4.3 Preliminary visual inspection ⎫ This is the main content
10.4.4 Detailed inspection (a long section) ⎭

10.4.5 Inspection of metallic linings ⎫ Specific requirements for various types of lined vessels
10.4.6 Inspection of non-metallic linings ⎭

There is a logic (of sorts) in the way this is set out. It attempts to be a chronological checklist of the way that a vessel is inspected. Don't forget the overall context of API 572 however; it is a technical *support* document for API 510 and so does not have to be absolutely complete in itself.

Overall, there is a lot of technical wisdom contained in API 572. The difficulty from an examination viewpoint is that it contains thousands of technical facts (and many opinions also) that can, theoretically, be chosen for exam questions. On the positive side, many API 572-related questions can be answered from general engineering inspection experience. You can improve your chances, however, by working through the code highlighting key points that may be you would *not* have anticipated from your experience.

Now try these familiarization questions.

6.8 API 572 section 10 familiarization questions

Q1. API 572 section 10.2.1: isolations

What kind of arrangement should be used to isolate a vessel?

(a) A thick steel plate held on by G-clamps ☐
(b) The plug-in types used for hydro tests ☐
(c) A proper ASME blank with the correct pressure/ temperature rating ☐
(d) A ring-flange with a rubber gasket ☐

API 572 Inspection of Pressure Vessels

Q2. API 572 section 10.2.1: other API codes
Which referenced API code deals with special precautions for entering vessels?

(a) API 579 ☐
(b) API 510 ☐
(c) API 2214 ☐
(d) API 2217A ☐

Q3. API 572 section 10.2.1: other API codes
Which referenced API code deals with sparking of hand tools?

(a) API 579 ☐
(b) API 2214 ☐
(c) API 2217A ☐
(d) API 660 ☐

Q4. API 572 section 10.2.1: gas tests
When should a gas test be done on a vessel?

(a) Only before the issue of the entry permit ☐
(b) Before the issue of the permit and after DP testing ☐
(c) Before the issue of the permit and periodically as required ☐
(d) Before the issue of the permit and before DP testing ☐

Q5. API 572 section 10.2.1: safety man (commonsense)
If an inspector feels faint when inside a vessel, what should the safety man do?

(a) Get further assistance ☐
(b) Do a gas test ☐
(c) Enter the vessel to help the inspector ☐
(d) All of the above ☐

Q6. API 572 section 10.3.2: ladders and walkways
What is wrong with doing a hammer test on bolts securing walkway plates?

(a) It can cause the bolt to fail by fatigue ☐
(b) It can cause the bolt to fail by brittle fracture ☐
(c) It can make the bolt come loose ☐
(d) Nothing ☐

Q7. API 572 section 10.3.3: foundations and supports

What is the situation with settlement of concrete vessel foundations?

(a) Any settlement at all is unacceptable ☐
(b) Settlement < 10 mm is acceptable ☐
(c) Settlement < 20 mm is acceptable ☐
(d) A nominal amount of even settlement is acceptable ☐

Q8. API 572 section 10.3.6: steel supports

What kind of distortion is most likely on vertical columns supporting a vessel?

(a) Stretching due to tensile stress ☐
(b) Shear due to compressive stress ☐
(c) Buckling ☐
(d) Torsion (twisting) ☐

Q9. API 572 section 10.3.6: steel supports

What causes corrosion on the inside of vessel skirts?

(a) High temperatures > 100 °C ☐
(b) Condensation ☐
(c) Galvanic cells ☐
(d) Increased stress owing to the vessel weight acting on the skirt ☐

Q10. API 572 section 10.3.8: nozzles

Which parts of a vessel nozzle assembly are at most risk of failure due to stresses imposed from misaligned pipework?

(a) The welds ☐
(b) The flange itself ☐
(c) The flange bolts ☐
(d) The nozzle parent material (due to hoop stress) ☐

SIMILAR SERVICE: FALLACY OR REASON?

To the dedicated follower of order and an easy life, the concept of similar service methodology provides rich pickings. The idea that identical vessels in similar service, experiencing similar process conditions will corrode in much the same way and at the same rate is all good news. There are fewer calculations to do and certainty feels more comfortable than doubt. Sadly there are critics of this approach citing, as they do, that on their site vessels under seemingly similar conditions corrode at vastly different rates, and some not at all. To reinforce the argument, they point to other similar things that differ in the world, such as pens that last for years, compared to others that dry up almost as soon as you have bought them.

Fortunately help is at hand, in the logical argument that proves once and for all that all vessels, anywhere, have precisely the same corrosion rate, whatever service conditions they see, how old they are or what they are made of.

Here's how. Suppose that we have a set of five vessels. We want to prove that they all have the same corrosion rate. Step back for a minute and suppose that we had a proof that all sets of *four* vessels has the same corrosion rate ... if that were true, we could prove that all five vessels have identical corrosion rates by removing a vessel to leave a group of four vessels. Do this in two ways and we have two different groups of four vessels. By our supposed existing proof, since these are groups of four, all vessels in them must have the same corrosion rate. For example, the first, second, third and fourth vessels constitute a group of four and thus must all have the same corrosion rate; and the second, third, fourth and fifth vessels also constitute a group of four and thus must also all have identical corrosion rates. For this to occur, all five vessels in the group of five must have the same corrosion rate (which is what we want to prove). Success beckons.

However, how do we know (as we assumed at the beginning) that all sets of four vessels have the same corrosion rate? Easy; just apply the same logic again. By the same process, a group of four vessels could be broken down into groups of *three*, and then a group of three vessels could be broken down into groups of two, and so on. Eventually you will reach a group size of one, and it is blindingly obvious, even to inspectors, that all vessels in a group of one must corrode identically, as there's only one of them.

Now the good news for larger sites ... by the same logic, the group size under consideration can also be increased. A group of five vessels can be *increased* to a group of six, and so on upwards, proving to corrosion engineers, one and all, that all finite-sized groups of vessels must have precisely the same corrosion rate; hence their services are no longer required.

So there you have it – proof, were it to be remotely required, of the subtle errors that can occur in attempts to conclude absolutely anything by induction.

Chapter 7

API 571 Damage Mechanisms

7.1 API 571 introduction

This chapter covers the contents of API 571: *Damage Mechanisms Affecting Fixed Equipment in the Refining Industry: 2003*. API 571 has only recently been added to the syllabus for the API 570 and 510 examinations and replaces what used to be included in an old group of documents dating from the 1960s entitled *IRE (Inspection of Refinery Equipment)*.

The first point to note is that the API 571 sections covered in the API 510 ICP exam syllabus are only an extract from the full version of API 571.

- Temper embrittlement
- Brittle fracture
- Thermal fatigue
- Erosion/corrosion
- Mechanical fatigue
- Atmospheric corrosion
- Corrosion under insulation (CUI)
- Cooling water corrosion
- Boiler condensate corrosion
- Sulphidation
- Chloride SCC
- Corrosion fatigue
- Caustic SCC/caustic embrittlement
- Wet H_2S damage (HIC/SOHIC)
- High-temperature hydrogen attack (HTHA)

Figure 7.1 The 15 damage mechanisms from API 571 in the API 510 exam syllabus

Group 1
Common DMs
- Brittle fracture
- Thermal fatigue
- Mechanical fatigue
- Corrosion fatigue

Group 2
Environment-related (more or less)
- Erosion/corrosion
- Atmospheric corrosion
- Corrosion under insulation (CUI)

Group 3
Higher temperature/more refinery-specific
- Boiler condensate corrosion
- Cooling water corrosion
- Sulphidation
- Chloride SCC
- Caustic SCC/embrittlement
- Wet H_2S damage (HIC/SOHIC)
- High-temperature hydrogen attack (HTHA)
- Temperature embrittlement

Figure 7.2　API 571 DMs revised order

7.1.1　The 15 damage mechanisms

Your API 510 exam copy of API 571 contains (among other things) descriptions of 15 damage mechanisms (we will refer to them as DMs). They are shown in Fig. 7.1.

Remember that these are all DMs that are found in the petrochemical/refining industry (because that is what API 571 is about), so they may or may not be found in other industries. Some, such as brittle fracture and fatigue, are commonly found in non-refinery plant whereas others, such as sulphidation, are not.

7.1.2　Are these DMs in some kind of precise logical order?

No, or if they are, it is difficult to see what it is. The list contains a mixture of high- and low-temperature DMs, some of which affect plain carbon steels more than alloy or stainless steels and vice versa. There are also several various

API 571 Damage Mechanisms

subdivisions and a bit of repetition thrown in for good measure. None of this is worth worrying about, as the order in which they appear is not important.

In order to make the DMs easier to remember you can think of them as being separated into three groups. There is no code-significance in this rearrangement at all; it is simply to make them easier to remember. Figure 7.2 shows the revised order.

One important feature of API 571 is that it describes each DM in some detail, with the text for each one subdivided into

REMEMBER THE WAY THAT API 571 COVERS EACH OF THE DAMAGE MECHANISMS

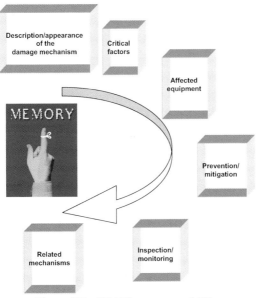

Figure 7.3   API 571 coverage of DMs

Caused by hydro-testing and/or operating below the Charpy impact transition temperature

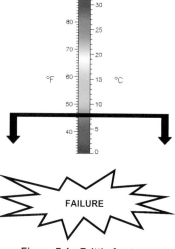

Figure 7.4 Brittle fracture

On a macro scale, thermal fatigue cracks tend to be dagger-shaped, wide and oxide-filled (caused by the oxidizing effect of the temperature variations)

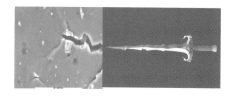

Joint restraint is a common cause of excessive thermal stresses

Figure 7.5 Thermal fatigue

API 571 Damage Mechanisms

six subsections. Figure 7.3 shows the subsections and the order in which they appear.

These six subsections are *important* as they form the subject matter from which the API examination questions are taken. As there are no calculations in API 571 and only a few graphs etc. of detailed information, you can expect most of the API examination questions to be *closed book*, i.e. a test of your understanding and short-term memory of the DMs. The questions could come from any of the six subsections shown in Fig. 7.3.

7.2 The first group of DMs

Figures 7.4 to 7.7 relate to the first group of DMs in API 571. When looking through these figures, try to cross-reference them to the content of the relevant sections of API 571. Then read the full sections of API 571 covering the four DMs in this first group.

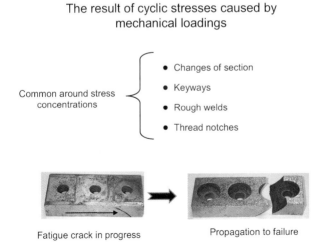

Figure 7.6   **Mechanical fatigue**

93

Quick Guide to API 510

| Cyclic loading | **+** | Corrosion |

= A type of fatigue cracking

Cracking is brittle, transgranular and not branched

Features of corrosion fatigue
- Often starts at pits and stress concentrations
- Can start at multiple sites
- Does not have a low stress limit
- May cause multiple parallel cracks

Figure 7.7   Corrosion fatigue

Attempt this first set of self-test questions covering the first group of API 571 DMs.

7.3 API 571 familiarization questions (set 1)

Q1. API 571 section 4.2.7.1: brittle fracture
Which of these is a description of brittle fracture?

(a) Sudden rapid fracture of a material with plastic deformation ☐
(b) Sudden rapid fracture of a material without plastic deformation ☐
(c) Unexpected failure as a result of cyclic stress ☐
(d) Fracture caused by reaction with sulphur compounds ☐

API 571 Damage Mechanisms

Q2. API 571 section 4.2.7.2: brittle fracture: affected materials

Which of these materials are particularly susceptible to brittle fracture?

(a) Plain carbon and high alloy steels ☐
(b) Plain carbon, low alloy and 300 series stainless steels ☐
(c) Plain carbon, low alloy and 400 series stainless steels ☐
(d) High-temperature resistant steels ☐

Q3. API 571 section 4.2.7.3: brittle fracture: critical factors

At what temperature is brittle fracture most likely to occur?

(a) Temperatures above 400 °C ☐
(b) Temperatures above the Charpy impact transition temperature ☐
(c) Temperatures below the Charpy impact transition temperature ☐
(d) In the range 20–110 °C ☐

Q4. API 571 section 4.2.7.4: brittle fracture

Which of these activities is *unlikely* to result in a high risk of brittle fracture?

(a) Repeated hydrotesting above the Charpy impact transition temperature ☐
(b) Initial hydrotesting at low ambient temperatures ☐
(c) Start-up of thick-walled vessels ☐
(d) Autorefrigeration events ☐

Q5. API 571 section 4.2.7.6: brittle fracture: prevention/mitigation

What type of material change will *reduce* the risk of brittle fracture?

(a) Use a material with lower toughness ☐
(b) Use a material with lower impact strength ☐
(c) Use a material with a higher ductility ☐
(d) Use a thicker material section ☐

Q6. API 571 section 4.2.7.5: brittle fracture: appearance

Cracks resulting from brittle fracture will most likely be predominantly:

(a) Branched ☐
(b) Straight and non-branching ☐
(c) Intergranular ☐
(d) Accompanied by localized necking around the crack ☐

Q7. API 571 section 4.2.9: thermal fatigue: description

What is thermal fatigue?

(a) The result of excessive temperatures ☐
(b) The result of temperature-induced corrosion ☐
(c) The result of cyclic stresses caused by temperature variations ☐
(d) The result of cyclic stresses caused by dynamic loadings ☐

Q8. API 571 section 4.2.9.3: thermal fatigue: critical factors

As a practical rule, thermal cracking may be caused by temperature swings of *approximately*:

(a) 200 °C ☐
(b) 200 °F ☐
(c) 100 °C ☐
(d) 100 °F ☐

Q9. API 571 section 4.2.9.5: thermal fatigue: appearance

Cracks resulting from thermal fatigue will most likely be predominantly:

(a) Straight and non-branching ☐
(b) Dagger-shaped ☐
(c) Intergranular ☐
(d) Straight and narrow ☐

Q10. API 571 section 4.2.9.6: prevention/mitigation

Thermal fatigue cracking is best avoided by:

(a) Better material selection ☐
(b) Control of design and operation ☐
(c) Better post-weld heat treatment (PWHT) ☐
(d) Reducing mechanical vibrations ☐

API 571 Damage Mechanisms

7.4 The second group of DMs

Figures 7.8 and 7.9 relate to the second group of DMs. Note how these DMs tend to be process environment-related. Remember to identify the six separate subsections in the text for each DM.

High fluid velocities cause scouring. Bends and welds are particularly susceptible

Notice the serious wall-thinning

TO REDUCE EROSION–CORROSION

- Reduce fluid velocity
- Use a more resistant material (harder alloys may be better)

Figure 7.8 Erosion/corrosion

CUI hides under lagging, and is often widespread

Chloride contamination (from water or lagging) makes CUI much worse

Figure 7.9 Corrosion under insulation (CUI)

7.5 API 571 familiarization questions (set 2)

Q1. API 571 section 4.2.14
A damage mechanism that is strongly influenced by fluid velocity and the corrosivity of the process fluid is known as:

(a) Mechanical fatigue
(b) Erosion–corrosion
(c) Dewpoint corrosion
(d) Boiler condensate corrosion

Q2. API 571 section 4.3.2: atmospheric corrosion
As a practical rule, atmospheric corrosion:

(a) Only occurs under insulation
(b) May be localized or general (widespread)
(c) Is generally localized
(d) Is generally widespread

Q3. API 571 section 4.3.2.3: atmospheric corrosion: critical factors
A typical atmospheric corrosion rate in mils (1 mil = 0.001 inch) per year (mpy) of steel in an inland location with moderate precipitation and humidity is:

(a) 1–3 mpy
(b) 5–10 mpy
(c) 10–20 mpy
(d) 50–100 mpy

Q4. API 571 section 4.3.3.3: CUI critical factors
Which of these metal temperature ranges will result in the most severe CUI?

(a) 0 to 51 °C
(b) 100 to 121 °C
(c) 0 to −10 °C
(d) 250+ °C

Q5. API 571 section 4.3.3.6: CUI appearance
Which other corrosion mechanism often accompanies CUI in 300 series stainless steels?

(a) HTHA
(b) Erosion–corrosion
(c) Dewpoint corrosion
(d) SCC

API 571 Damage Mechanisms

7.6 The third group of DMs

Now look through Figs 7.10 to 7.15 covering the final group of DMs. These DMs tend to be either more common at higher temperatures or a little more specific to refinery equipment than those in the previous two groups.

Again, remember to identify the six separate subsections in the text for each DM, trying to anticipate the type of examination questions that could result from the content.

This is a high-temperature corrosion mechanism

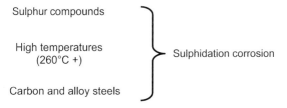

The main problem is caused by H_2S (formed by the degradation of sulphur compounds at high temperature)

Occurs in crude plant, cokers, hydroprocessor units, fired heaters, etc. – anywhere where there are high-temperature sulphur streams

Figure 7.10   Sulphidation corrosion

Quick Guide to API 510

API terminology calls it 'Environmental-*assisted cracking*'

The stress exposes the grain boundaries to corrosion

Temperature range above 60°C (140 °F) and pH > 2

- One of the most common corrosion mechanisms
- Prevalent in 300 series austenitic stainless steel and high-chromium alloys

Figure 7.11   Stress corrosion cracking (SCC)

API 571 Damage Mechanisms

A specialist type of stress corrosion cracking caused by alkaline conditions. The worst offenders are:

- Caustic potash (KOH)

- Sodium hydroxide (NaOH)

Caustic attack in a heat exchanger tubesheet

Typically found in H_2S removal units and acid neutralization units

Figure 7.12 Caustic embrittlement

This is a specialist and complex corrosion mechanism

At high temperatures, H_2 reacts with the carbon in the steel forming CH_4 (methane)

The resulting loss of carbides weakens the steel

Fissures start to form and propagate into cracks

Figure 7.13 High-temperature hydrogen attack (HTHA)

Quick Guide to API 510

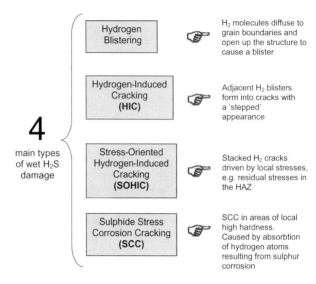

All result in blistering or cracking of low carbon and low alloy steels

These wet H_2S DMs have quite complex descriptions
and critical factors. Please refer to API 571 (5.1.2.3)

Figure 7.14   Wet H_2S damage

API 571 Damage Mechanisms

This is a specialist type of brittle fracture, not related to low temperatures

650–1000°F
Long-term exposure

Refinery reactors, separators, etc.)

Low-alloy steels
(e.g. 2.25Cr–1Mo)

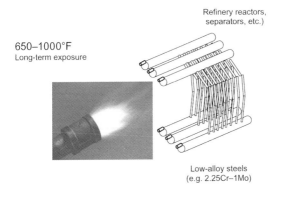

The high temperature exposure causes the material to become more brittle (lower Charpy toughness) over time. See API 571(4.2.3)

Brittle fracture by cracking

Typical flat fracture face

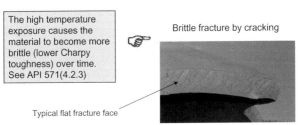

Figure 7.15   Temper embrittlement

POSTSCRIPT:
THE CASE FOR A CORROSION RATE (OF SORTS)

To the casual reader of codes, handbooks and suchlike the world of corrosion looks a more or less rational place. Pressure vessels enter service amidst a modest fanfare of applause and back-slapping and then happily settle down to corrode quietly away. Time passes, inspectors come and go, and vessels are retired with grace, precisely at the end of their useful life, just before they become dangerous. Governing all of this is the existence of a known and uniform corrosion rate; the inevitable result of one of more of the damage mechanisms (DMs) paraded, eloquently and in sequence, in API 571. Displayed together this represents a tidy package of fact and evidence that, it is argued, never harmed anyone.

Fact comes into it in the premise that this corrosion rate is actually *based on fact*, i.e. that the DMs that is producing it has actually been found and identified beyond all absolute doubt – a task somewhat akin to finding a needle in a haystack. Like needles, DMs have a nasty habit of hiding themselves away precisely where they are difficult to find. It is as if they know that you will have difficulty removing all the lagging or vessel internal fittings, so off they go to hide behind them.

The *evidence* part is pretty clear. The main technical character of DMs is their almost guaranteed *un*predictability. Vessel degradation and failures in many different industries worldwide stand in evidence of this; most unexpected failures are caused by either fatigue, some complex and largely unpredictable cocktail of DMs that would have been next to impossible to predict, or a combination of the two.

However, before we run off with the idea that, when talking about a uniform corrosion rate, both faith and doubt are equally misplaced, remember that in order to set a remaining lifetime and inspection period for a corroding vessel, you have to start *somewhere*.

Maybe that's what API 510 exam corrosion rate questions are about.

Chapter 8

API 576 Inspection of Pressure-Relieving Devices

8.1 Introduction to API 576

This chapter is about learning to become familiar with the layout and contents of API 576: *Inspection of Pressure-Relieving Devices*. Similar to API 572, API 576 is a well-established document (it is still on its 2000 edition) with its roots in earlier documents published by the American refining industry. It is more a technical *guide document* rather than a code, as such, but it does perform a useful function in supporting the content of API 510.

Note the following four points about API 576:

Point 1. It is a *well-detailed and comprehensive technical document*. Unlike some API codes, which have a very selective approach to their subject, API 576 is one of the best in the quality of the technical information it provides. It is an excellent guide to the practical aspects of pressure-relieving devices.

Point 2. API 576 introduces *specific API terminology* on the types of pressure-relieving devices. These relate to the definitions used for the following terms: safety valve, relief valve and safety relief valve. These API definitions are technically consistent in themselves but can be different to those used in other codes (e.g. BS/EN/DIN).

Point 3. Similar to API 572, it refers to a few *related codes* that are not in the API 510 exam syllabus (mainly API 527 covering seat leakage testing; see API 576 section 2 on page 1 of the code). As with previous related codes, you need to know that these additional codes exist but you do not need to study them for the API 510 examination.

And finally, the most important point. Like API 572, API 576 *is all text and technical descriptions*, accompanied by explanatory photographs of a fairly general nature. It contains no calculations. This means that many examination questions about API 576 in the API 510 certification exam may be *closed book*. The downside to this is that API 576 contains many separate technical facts, giving a large scope for the choice of exam questions.

Again, similar to API 572, appendix A of API 576 contains specimen work order/inspection report sheet/test record formats. While these undoubtedly contain sound guidance on the format of reports they are not a particularly suitable subject matter for multiple choice exam questions.

All this means that you need to concentrate firmly on the *technical* (rather than administrative) content of API 576. We will look at some of the more important areas as we work through the code.

8.2 API 576 sections 3 and 4: types (definitions) of pressure-relieving devices

Remember that API 576 uses *specific definitions* for the various types of pressure-relieving devices and that these may not correspond with those in other codes (or your own knowledge and experience). API 576 sees the situation as shown in Fig. 8.1.

Don't worry if you find this a bit confusing. Read the following points, which attempt to clarify the situation:

NOTE THESE POINTS ABOUT PRV TERMINOLOGY

- Note the first annotation on Fig. 8.1; it shows the term PRV as a generic definition covering most practical types of pressure-relieving device, excluding bursting discs. We will use this term **PRV** in this context throughout the rest of this chapter.
- Now read API 576 section 4.2: safety valve. Note how the 'pure' type of safety valve described refers, in the main, to valves used for *steam* service (it doesn't actually say this, but that is what it means).

API 576 Inspection of Pressure-Relieving Devices

API 576 PRV TERMINOLGY

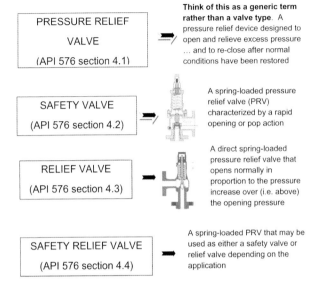

Figure 8.1 **Pressure-relieving device definitions**

- Now read API 576 section 4.3: relief valve. Note how the 'pure' type of relief valve described refers, in the main, to valves used for *liquid* service (again, it doesn't actually say this, but this is what it means).
- Now read API 576 section 4.4: safety relief valve. Can you see why this has been introduced? Think of it as a generic definition, encompassing both the 'pure' types of *safety valve* and *relief valve*. The reason for API 576 introducing this term *safety relief valve* is to take into account the fact that many proprietary designs of PRV can be used on gas/vapour *or* liquid service and so can be considered (with a bit of imagination) as both a safety valve *and* a relief valve,

depending on application. Now you see why it is important not to confuse the terminology.

Keep these definitions in mind as you work through the rest of this API 576 chapter and they should start to become clearer.

8.3 Types of pressure-relieving device

Section 4 of API 576 describes the different types of pressure-relieving device. Note how this is supplemented by the various terminology definitions in section 3.

Protective device types: API 576

There are a large number of different types of protective device.

Spring-loaded valves

- Normal relief valves
- Safety valves
- Balanced safety relief valves (bellows)
- Pilot operated relief valves

Rupture discs (not re-useable)

- Conventional
- Scored tension
- Reverse acting
- Graphite

Vacuum valves

- Dead weight (more often seen on tanks)
- Pilot operated valves (diaphragm)
- Spring and dead weight

As we are mostly concerned with internal pressure, we shall concern ourselves mostly with spring-loaded safety valves and rupture discs. Before we look at protective devices, we will look at the various internal pressures that can exist inside a vessel. These are summarized in Fig. 8.2.

API 576 Inspection of Pressure-Relieving Devices

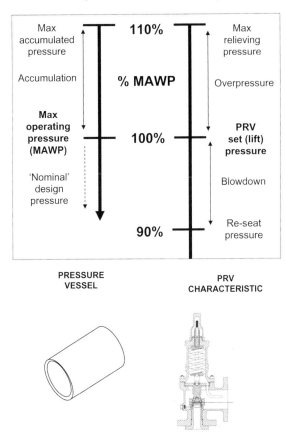

Figure 8.2 Pressure term definitions

Normal maximum operating pressure. This is a bit confusing. It is, essentially, the design pressure based on the ASME VIII code.

Maximum allowable working pressure (MAWP). This is equal to or greater than the nominal design pressure (it is based, again, on ASME VIII and a common calculation parameter included in the API 510 examination syllabus).

Accumulation. This is the pressure *above* MAWP or design pressure (values defined in ASME VIII).

Overpressure. This may be expressed as either:

- The pressure above the set pressure of a PRV at which full discharge flow is achieved (mass flow kg/s or lb/s).
- A percentage of set pressure.
- The same as accumulation (when based on MAWP).

Note that the actual **set pressure** of the valve or disc is vital to the performance of the vessel. If it is set too low (at or just above the operating pressure) the valve will lift frequently. If it is set too high (above MAWP) the vessel may become over-pressurized. There are a lot of terms for the different operational characteristics of a PRV, which are given in API 576 section 3.4. Two terms that deserve further explanation are *back pressure* and *cold differential test pressure (CDTP)*:

- **Back pressure** is the pressure that is present downstream of the discharge pipe of the PRV. It can either be present all the time or can build up as the result of flow as the PRV opens. For example, if a valve has a set pressure of 100 psi and is vented into a system that is operating at 25 psi, the valve will not operate until 125 psi.
- **Cold differential test pressure (CDTP)** is used for PRVs on 'hot' duty and is an adjusted pressure at which the valve opens on a test stand at room temperature. The CDTP is corrected for both temperature and back pressure.

API 576 section 4: pressure-relieving valve types

Section 4.3: relief valves
Relief valves (see Fig. 8.3) are direct spring-loaded valves that begin to open when the set pressure is reached. They open progressively and do not exhibit a pop action. Full lift is

obtained with an overpressure of *10 % or 25 %* depending on the type of valve. The valve will close after blowdown is complete but at a pressure *lower* than the set pressure. They are sometimes called thermal relief valves as they may relieve small pressure increases caused by thermal expansion of process liquid.

Normally, relief valves have a closed bonnet to prevent the release of product that is toxic, corrosive, flammable or expensive. For tightness of the seat, resilient O-rings can be fitted to replace the conventional metal-to-metal seat.

Applications: mostly used on incompressible liquids.

Limitations: relief valves should *not* be used on the following services:

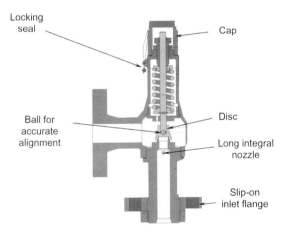

The term 'thermal' relief valve is sometimes used for this type of PRV. It simply indicates that it may be thermal expansion of system liquid that causes the PRV to open to relieve the pressure, rather than any serious upset condition

Figure 8.3   A thermal relief valve

- Vapour services including steam, air and gas.
- Systems that discharge into a closed header unless the back pressure build-up has been allowed for.
- As a bypass or pressure control valve.

Section 4.7: pilot-operated pressure relief valves

In this type, an auxiliary pressure relief valve called the *pilot* actuates the main valve. The pilot may be mounted on the same connection as the main valve or separately. The pilot valve opens normally at the set pressure and in turn operates the main valve. Figure 8.4 shows the principle.

Applications:

- Used when a high set pressure and large relief area are required. This type of PRV can be set to a maximum flange rating.
- The differential pressure between the operating and set pressures is small.
- Large low-pressure storage tanks.
- Where very short blowdown is necessary
- To replace bellows type valves to overcome the problems of high back pressure.
- Where pressure can be sensed in one position and the process is relieved at another.
- Where frictional pressure losses in the inlet or outlet pipework are high.

Limitations: Pilot PRVs have limitations on systems where:

- Service duty is dirty (unless filters are fitted in the system).
- High viscosity fluids (small passages in the pilot valve slow down the flow rate).
- The process fluids may form polymers, which cause blockages.
- High temperatures where seals, O-rings or diaphragms may not be suitable.
- The process may attack the seals, O-rings or diaphragms.
- Corrosion build-up affects the operation of the pilot valve.

API 576 Inspection of Pressure-Relieving Devices

The pilot valve senses the system pressure. When it lifts, it vents the main valve dome and the main valve piston lifts under system pressure, opening the valve

Pilot valve

Main valve

Dome

Full bore discharge

Piston

System pressure inlet

Sensing pipe

These are also sometimes called 'full bore' PRVs as they allow full bore discharge of the process fluid

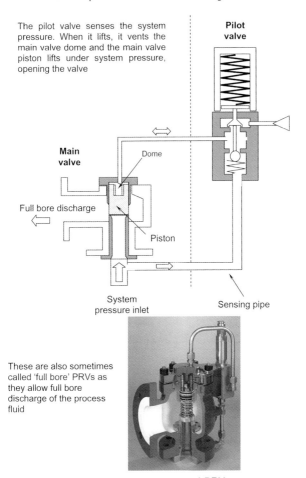

Figure 8.4   Pilot operated PRV

Rupture disc devices

Rupture (or bursting) discs comprise a thin disc of material that has a known bursting pressure, held within a special holder. The actual disc can be made in a number of different configurations (see Fig. 8.5):

- Domed
- Reverse acting
- Flat

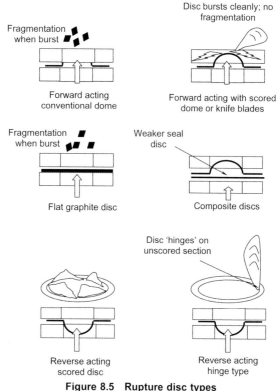

Figure 8.5 Rupture disc types

Conventional (often called *direct acting*) **discs**. The pressure acts on the concave side of the disc.

Scored tension loaded rupture disc. The pressure acts on the concave side but the disc is cut or scored by mechanical means during fabrication and is made of thicker material. The objective is to achieve a more accurate, reliable burst pressure.

Composite rupture disc. These are flat or domed, made in metallic or non-metallic material, with multipiece construction. The top section is slit and the burst pressure is controlled by the combination of the top and the underlying sections.

Reverse-acting rupture disc. These are domed but the pressure is on the convex side. The disc can be ruptured by a number of methods:

- Shear
- Knife blades
- Knife rings
- Score lines

Graphite rupture disc. A flat disc of graphite impregnated with a binder material. It bursts by bending or shear.

8.4 API 576 section 5: causes of improper performance

This section of API 576 contains reasons for pressure-relieving devices (mainly PRVs) failing to work properly. The section is not particularly well structured but does contain good technical details. Much of the section is taken up with the corrosion/damage mechanisms that affect PRVs. Here is a summary of the content.

Section 5.1: corrosion
Because PRVs are on the same duty as the pressure vessels they protect, they are subject to the many causes of corrosion

that also occur in the vessels. Some typical (refinery industry) examples given in section 5 are:

- Acid attack on carbon steel due to a leaking valve seat
- Acid attack on a stainless steel inlet nozzle
- Chloride corrosion on a stainless steel nozzle
- Sulphide corrosion on a carbon steel disc
- Chloride corrosion on a stainless steel disc
- Pitting corrosion on stainless steel bellows
- Sour gas (H_2S) attack on a monel rupture disc

By careful selection of the correct materials within a PRV, most corrosion problems can be overcome. The correct maintenance of the valve also stops any potential leakage allowing corrosive processes into those parts that could rapidly deteriorate.

As previously mentioned, *bellows* are used to give protection to the valve spring and discharge side of the valve. Also, placing a rupture disc directly under a PRV gives added protection to the valve components.

Section 5.2: damaged seating surfaces
Because there is metal-to-metal contact between the valve disc and nozzle, this area must be extremely flat, as any imperfections will lead to a process leak. The seating surfaces must be lapped to produce a finish of two to three light bands (0.000 034 8 in). Figure 8.6 shows the principle.

Section 5.3: failed springs
Safety valve springs fail in two distinct ways:

Gradual weakening, which can cause the valve to open 'light'. This may be caused by:

- Improper material choice for the spring
- Operating at temperatures too high for the material
- Corrosion, leading to cracks and failure

API 576 Inspection of Pressure-Relieving Devices

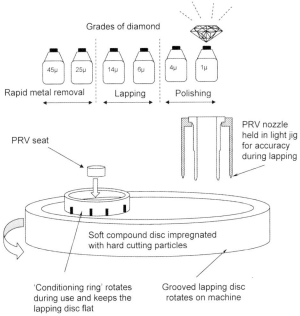

Figure 8.6 **PRV seat lapping**

Catastrophic failure of the spring, so that the valve opens and jams, preventing it from closing. The normal cause of this type of failure is stress corrosion cracking (SCC).

Section 5.4: improper setting and adjustment
Several factors can affect the setting and adjustment of a PRV:

- Not following the manufacturer's instructions.
- Testing the valve with the wrong medium. Water, air or nitrogen is frequently used:
  - Gases generally produce a definite 'pop' and are generally used for vapour service.
  - Water is generally used for liquid service.

- Steam service should be tested with steam, to replicate its temperature and flow characteristics.
- Incorrect pressure gauge calibration. Gauges should be tested with a calibrated dead weight tester. The test pressure should fall within the middle third of the test gauge.
- Blowdown rings not correctly set.

Section 5.5: plugging and sticking
When working on a fouling fluid, the inlet pipe to a PRV can become completely blocked. The fouling can be caused by a large number of refining industry processes that give solid particles such as coke and iron sulphide.

Section 5.6: misapplication of materials
Occasionally, the material of construction of a PRV is not suitable for the process duty. Hydrogen sulphide and chloride attack are typical examples.

Section 5.7: improper location, history or identification
A valve may not provide the required protection if is not located at the correct location. A record should be maintained of the history of the valve including the specification, any repairs, installation details, etc.

Section 5.8: rough handling
PRVs are manufactured and maintained to a commercial seat tightness standard given in API 527. Rough handling may change the set pressure or otherwise cause damage to the valve.

Section 5.9: improper differential between operating and set pressures
In use, a PRV should be kept tightly closed by having a reasonable margin of difference between the operating and set pressures. The design of the system governs the operating and set pressures, and references to the guidelines are found in ASME VIII.

API 576 Inspection of Pressure-Relieving Devices

8.5 API 576 section 6: inspection and testing

This section contains the main inspection and test activities traditionally used on PRVs. It describes (in roughly chronological order) the stages shown in Fig. 8.7.

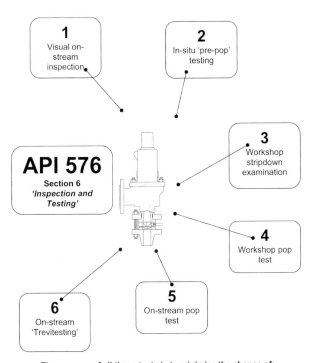

The purpose of all these tests is to minimize the chance of:

- Failure of the pressure boundary (or 'envelope')
- Failure to lift at the correct pressure
- Leakage across the seat in service
- Failure to reseat after lifting

Figure 8.7 PRV inspection and test activities

8.6 API 576 familiarization questions

Q1. API 576 section 3.3.1: system pressures
The amount by which the pressure in a vessel rises above MAWP when the pressure-relieving device is fully open and discharging is known as?

(a) Back pressure build-up
(b) Accumulation
(c) CDTP
(d) Blowdown

Q2. API 576 section 3.3.3: system pressures
Generally speaking, which of the following is true for a pressure vessel (according to API 576)?

(a) MAWP is always the same as design pressure
(b) MAWP is normally greater than design pressure
(c) MAWP is normally lower than design pressure
(d) MAWP = design pressure – back pressure

Q3. API 576 section 3.4: device pressures
What is back pressure?

(a) Pressure in the PRV inlet line
(b) Pressure in the PRV discharge line before the PRV lifts
(c) Pressure in the PRV discharge line after the PRV lifts
(d) The sum of (b) and (c) above

Q4. API 576 section 3.4.7: CDTP
Cold differential test pressure (CDTP) is:

(a) The pressure at which a PRV is set to lift on the test stand
(b) The pressure at which a PRV is set to lift under service (hot) conditions
(c) 'Set pressure' + 10 %
(d) 'Set pressure' + back pressure

Q5. API 576 section 4.1: pressure relief valves (PRV)
In API 576, the term 'PRV' refers to:

(a) Only valves that exhibit a defined 'pop' action
(b) Only valves that do not exhibit a defined 'pop' action
(c) Only those valves defined as 'safety' relief valves (API 576 section 4.4)
(d) All of the above. PRV is a generic term

API 576 Inspection of Pressure-Relieving Devices

Q6 API 576 section 4.2.2: safety valve limitations
Can safety valves be used in corrosive service without being isolated from the process fluid by a rupture disc on the inlet side?

(a) No, it is not recommended ☐
(b) Yes, as long as the seat is corrosion-resistant ☐
(c) Yes, as long as it has an enclosed bonnet ☐
(d) Yes, as long as the fluid is compressible ☐

Q7. API 576 section 4.3: relief valve
What is the fundamental difference between the opening characteristic of a relief valve compared to that of a safety valve?

(a) A relief valve has a lower measured lift ☐
(b) A relief valve opens without a 'pop', in proportion to the pressure increase over the opening pressure ☐
(c) A relief valve contains a huddling chamber, giving a proportional opening ☐
(d) A relief valve should not be used on liquids ☐

Q8. API 576 section 4.9.3: rupture disc limitations
Which of the following damage mechanisms (DMs) would be *unlikely* to affect conventional rupture discs, causing premature failure?

(a) Fatigue ☐
(b) Stress corrosion cracking ☐
(c) Creep stress failure ☐
(d) Brittle fracture ☐

Q9. API 576 section 5.3: failed springs
Failed PRV springs are almost always caused by:

(a) Plastic deformation ('set') of the spring due to continual use ☐
(b) Fatigue ☐
(c) Brittle fracture ☐
(d) Surface corrosion and/or stress corrosion cracking ☐

Q10. API 576 section 5.5: plugging and sticking
Which of the following is *unlikely* to be a cause of sticking of a valve disc in its guide?

(a) Machining of components outside their tolerance limits ☐
(b) Use on process fluids such as coke or catalysts ☐
(c) Use of a balanced (bellows-type) PRV ☐

(d) Galling of mating components

Q11. API 576 section 5.8: PRV handling
Rough handling of a PRV can result in problems with seat leakage. The standard used to specify PRV leakage is?

(a) API 576
(b) API 596
(c) API 527
(d) API 572

Q12. API 576 section 5.8.1: handling during shipment
In order to minimize the chances of damage to PRV seating surfaces, PRVs should be shipped:

(a) In an upright position
(b) Lying on their side (firmly secured to a pallet)
(c) With the spring fully compressed
(d) With the spring fully extended

Q13. API 576 section 6.2.1: safety aspects of PRV inspection
Before removing PRVs from the plant, it is an important safety requirement to check that:

(a) The spring is released
(b) The connecting pipework and any block valves are adequately supported
(c) They are 'pre-pop' tested first
(d) The discharge connection to atmosphere is blanked off

Q14. API 576 section 6.2.8: as-received pop pressure
During its first as-received pop test, a PRV opens at 120 % CDTP. It is tested a second time and opens at 105 % CDTP (a pressure considered acceptable under the applicable code). Which pop pressure result should be used as the basis of determining the inspection interval for this PRV?

(a) 105 % CDTP
(b) 105 % CDTP ± 10%
(c) 120 % CDTP
(d) CDTP

Q15. API 576 section 6.2.8: 'as-received' pop pressure

When is it acceptable for a user to waive the 'as-received' pop test on a very dirty PRV and still be in compliance with API 576?

(a) Never ☐
(b) If the PRV is the balanced (bellows) type ☐
(c) If the PRV has not been in HF service ☐
(d) If the inspection interval is immediately reduced, and then assessed again at the next inspection ☐

Chapter 9

ASME VIII Pressure Design

9.1 The role of ASME VIII in the API 510 syllabus

ASME VIII-I *Rules for the Construction of Pressure Vessels* is a long-established part of the API 510 syllabus (or 'body of knowledge' as they call it). Strictly, it is concerned only with the new construction of unfired vessels (i.e. not boilers or fired heat exchangers) and so contains no reference in it at all to what happens to vessels after they are put into service.

Now you see the problem; the API 510 syllabus is almost the antidote to ASME VIII in that its role is to give guidance of what to do with vessels that have been in service for some time. In doing this it has to deal with vessels that may be corroded, have been re-rated or in some way are not the same as when they were new.

Fortunately, the answer is fairly straightforward and centres around the situation when a vessel is being repaired or altered; the title of API 510 is *Inspection, Repair and Alteration* remember. API 510 sees vessel repairs and alterations as a straightforward remanufacturing exercise that should follow the requirements of ASME VIII-I, just as when it was built. The added difficulty of doing this on-site rather than in the manufacturing shop brings two implications:

- An API 510 qualified inspector has to know some relevant points of ASME VIII.
- There may be some areas in which ASME VIII either physically cannot be followed (because the vessel is already built) or in which an alternative, perhaps less conservative, solution is adequate.

Figure 9.1 shows the solution. Note how the blanket coverage of repairs by ASME VIII is overridden in a few well-chosen areas written into API 510.

ASME VIII Pressure Design

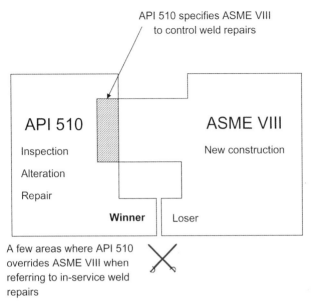

Figure 9.1 The situation with ASME VIII

9.2 How much of ASME VIII is in the API 510 syllabus?

ASME VIII is a complex multilayered network of intrigue. It is all held together by a rather confusing system of clause numbering, littered with a nested interlinking network of cross-references. It is no doubt a proven, competent code, but it can look confusing if you don't deal with it regularly or, even worse, have never seen it before.

The good news is that *not much* of the content of ASME VIII is included in the API 510 syllabus (formally called the 'body of knowledge' remember). You can think of it, simplistically, as involving only a few minor abstracts plucked out of the code and used as the subject of a small family of examination questions. Look at it carefully and you will conclude that the scope of these questions is less, and the content is more straightforward, than the API body of knowledge infers.

Practically, the subjects chosen from ASME VIII relate in some way to the repair or replacement of vessel components. This makes sense, as this is what API 510 is all about. While the individual topics are not difficult, the way in which they fit together is important to understand. Figure 9.2 shows the situation – note how all the individual topics are not linear, as such, but act together to specify the 'design' of a repaired

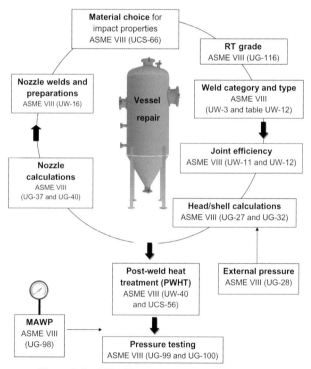

Figure 9.2　API 510 syllabus: ASME VIII content

ASME VIII Pressure Design

or replaced pressure vessel component. For simplicity, we will look at each of these in turn.

9.2.1 Material choice

The API 510 syllabus takes a fairly one-dimensional view of material choice. Unlike the API 570 syllabus, which utilizes ASME B31.3, ASME VIII contains no material data at all. Instead, it references the full data set of ASME II (d), which thankfully is not in the API 510 syllabus. The main material property that API 510/ASME VIII is concerned with is that of resistance to brittle fracture. The fundamental issue is therefore whether a material is suitable for the minimum design metal temperature (MDMT) for which a vessel is designed. This topic is covered by clause UCS-66 of ASME VIII.

9.2.2 Vessel design features

The main ASME VIII design topics required included in the API 510 syllabus are:

- Internal pressure in shells and heads (clauses UG-27 and UG-32)
- External pressure on shells (clause UG-28)
- Nozzle compensation (mainly figure UG-35.1)
- Nozzle weld sizing (mainly figure UW-16)

All four of these topics are heavily and artificially simplified for the purpose of producing API 510 exam questions. Without this simplification, calculation questions would be just too complex for a 4-minute exam answer.

9.2.3 RT grades

ASME VIII has a fairly unique approach to vessel design in that for every specified vessel 'design requirement' there is not just one but up to four possible design solutions. Put another way, if you specify an ASME VIII-I vessel for a specific set of pressure–temperature criteria you could legitimately receive up to four different 'designs', all of which are fully compliant with the specific requirements of the code.

The difference comes from the concept of 'RT grade'. This works off three simple principles:

- ASME VIII allows a degree of freedom in the amount (scope) of NDE done on a vessel.
- The amount of NDE that is actually chosen is denoted by the RT grade: RT-1, RT-2, RT-3 or RT-4, as set out in clause UG-116 of ASME VIII-I.
- The RT grade that is chosen defines the joint efficiency E, which is used for the shell and head design calculations. A high joint efficiency of 1 (when grade RT-1 is chosen) gives the smallest wall thickness.

9.2.3.1 RT grade terminology

This is not the easiest concept of ASME VIII to understand on first reading. Figure 9.3 shows the situation. Note how both RT-1 and RT-2 can be referred to as 'full' radiography grades, even though strictly they are not. It is easier to just accept this, rather than look too deeply into the logic behind it. Grade RT-3 is referred to as the 'spot' radiography grade whereas RT-4 means 'less than spot RT', which also includes no RT at all.

As you can see from Fig. 9.3, the most awkward grade to understand is RT-2. We will explain this later in Chapter 10; for the moment just think of it as a 'full' RT category.

9.3 ASME VIII clause numbering

ASME VIII uses a system of clause numbering that can look confusing when you first encounter it. The full code is divided into multiple sections designated by letters (UG, UW, UCS, UHT, UF, UB, UNF, UCI and similar). Most of these are not required for the API 510 syllabus; the only ones you actually need *parts of* are as follows.

- UG: the G denotes *general* requirements.
- UW: the W denotes *welding*.
- UCS: the CS denotes *carbon steel* (so you can expect this

THE RT GRADES OF UG-116 (simplified)

RT-1 'Full' designation indicates:

- 100 % RT of all longitudinal and circumferential seams
- 100 % RT of Cat B and C only nozzle welds over 10 in diameter and of weld neck design (this does not include any of the Cat D nozzle-to-shell welds)
- Basic $E = 1$

RT-2 'Full' ('Spot Plus') designation indicates that the extra radiograph of UW-11 (a)(5)(b) has been complied with, so

- 100 % RT of longitudinal weld seams
- Spot RT of circumferential seams (plus the extra one)
- No RT done on Cat B and C welds in **any** nozzles
- Basic $E = 1$

RT-3 'Spot' designation indicates that the extra radiograph of UW-11 (a)(5)(b) *has not* been complied with, so

- Spot RT on all longitudinal and circumferential seams
- No RT done on **any** nozzles
- Basic $E = 0.85$

RT-4 designation indicates:

Any amount of RT that does not comply with RT-1, RT-2 or RT-3

Figure 9.3 ASME VIII RT grades

section to be concerned with materials of construction and their heat treatment).

- UHT: the HT denotes *heat treatment* (this part contains specific requirements for heat-treated ferritic steels).
- Appendix 1: supplementary design formulae
- Appendix 3: definitions

Remember that not all of the ASME VIII code pages provided in your API 510 document package are actually needed. Only specific paragraphs are in the examination scope. Of the above, the majority of the technical details relating to welding are contained in section UW. A few others appear in section UCS.

We will now look at one of the major topics of the UG section of ASME VIII. This section deals predominantly with design. The first topic to be covered (internal pressure loadings) acts as an introduction to the design concepts of ASME VIII. Subsequent sections look at the related subjects of external pressure loadings, nozzle compensation, pressure testing and materials issues such as impact test and heat treatment.

9.4 Shell calculations: internal pressure

Shell calculations are fairly straightforward and are set out in UG-27. Figure 9.4 shows the two main stresses existing in a thin-walled vessel shell.

Hoop (circumferential) stress

This is the stress trying to split the vessel open along its length. Confusingly, this acts on the *longitudinal* weld seam (if there is one). For the purpose of the API 510 exam this is the governing stress in a shell cylinder. The relevant UG-27 equations are:

$$t = \frac{PR_i}{SE - 0.6P}$$

(used when you want to find t) or, rearranging the equation to find P when t is already known:

$$P = \frac{SEt}{R_i + 0.6t}$$

where
- P = maximum design pressure (or MAWP).
- t = minimum required thickness to resist the stress.

ASME VIII Pressure Design

For API 510 exam purposes, the important stress for calculation purposes is the **circumferential (hoop) stress** as this is the one that governs the required shell thickness

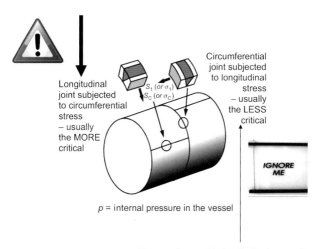

Figure 9.4 Vessel stresses

These only apply to thin-walled shells. That is all the API 510 exam questions normally cover

- S = allowable stress of the material. This is read from ASME IID tables or, more commonly, given in the exam question (it has to be as ASME IID is not in the syllabus).
- E = joint efficiency. This is a factor (between 0.65 and 1) used to allow for the fact that a welded joint may be weaker than the parent material. It is either read off tables (see UW-11 and UW-12 later) or given in the exam question. You can think of E as a safety factor if you wish.

R_i = the internal radius of the vessel. Unlike some other design codes ASME VIII-I prefers to use the internal radius as its reference dimension, perhaps because it is easier to measure.

A key feature of R_i is that it is the radius in the corroded conditions (i.e. that anticipated at the next scheduled inspection). Don't get confused by this – it is just worked out in this way. If a vessel has a current R_i of 10 in and has a corrosion rate (internal) of 0.1 in/years, with the next scheduled inspection in five years, then:

Current R_i = 10 in
R_i in 5 years = 10 in + (5 × 0.1 in) = 10.5 in corroded condition

Hence 10.5 in is the R_i dimension to use in the UG-27 equation.

Axial (longitudinal) stress

This is the stress trying to split the vessel in a *circumferential* plane; i.e. trying to pop the head off the shell. It is approximately half the magnitude of the hoop stress and so not a 'governing' design parameter (at least for the purpose of the API 510 exam). You can see that the equation for it is in UG-27 – but probably they rarely, if ever, appear in exam questions. Now note the following specific points.

Looking at the formula for the cylindrical shell in UG-27 (c) (1) for circumferential stress, there are two limitations applied to it:

- The thickness must not exceed one-half of the inside radius, i.e. it is not a thick cylinder.
- The pressure must not exceed $0.385SE$, i.e. not be high pressure. In practice this is more than about 4000 psi for most carbon steel vessels.

ASME VIII Pressure Design

If the first requirement applies, i.e. you are dealing with a thick cylinder, then a completely different set of equations is needed. Don't worry about them – they won't be in the exam.

9.4.1 Shell calculation example
The following information is given in the question.

R = inside radius of 30 in (R_i)
P = pressure of 250 psi (MAWP)
E = 0.85 (type 1 butt weld with spot examination as per UW-12)
S = 15 800 psi

What minimum shell thickness (which may be called t_{min} or $t_{required}$) is necessary to resist the internal MAWP?

Using thickness $(t) = PR/(SE-0.6P)$ from UG-27
Thickness = $250 \times 30 / [15800 \times 0.85 - (0.6 \times 250)]$

t = 0.565 in   ANSWER

Remember that this is exclusive of any corrosion allowance that you decide to add.

Now, as an exercise, look at section UG-16 (c) covering mill undertolerance. Note how, strictly, you are allowed an undertolerance of 0.25 mm or 6 % of design thickness while still using the design pressure. Don't worry about this because if exam questions require you to use this undertolerance allowance, they will mention it in the questions.

Now work the equation from the other perspective, where you are given a shell material thickness and you have to determine the maximum acceptable pressure (MAWP):

Given t = 0.625 in
Using pressure $(P) = SEt/(R + 0.6t)$ from UG-27
Pressure (P) = $15800 \times 0.85 \times 0.625 /$
$[30 + (0.6 \times 0.625)]$

MAWP = 276 psi   ANSWER

9.5 Head calculations: internal pressure

9.5.1 Head thickness

Pressure vessels have the pressure enclosed by a head that has the pressure acting on the inside of the head. There are a number of different types of heads that can be used:

- Ellipsoidal
- Torispherical
- Hemispherical
- Conical

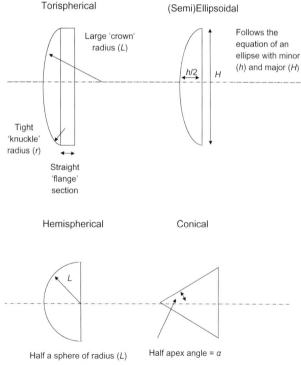

Figure 9.5 Vessel head shapes

ASME VIII Pressure Design

- Toriconical

These are all in the API 510 syllabus except for the toriconical design. Figure 9.5 shows the shapes. The formulae for the required minimum thickness are given in ASME VIII UG-32 (d) to (g).

9.5.2 Ellipsoidal heads UG-32 (d)

The ratio of major to minor axis is set at 2:1 for a standard ellipsoidal head. This is the type assumed by the UG-32 (d) formula (see Fig. 9.6). When the ratio of the major to minor axis is not 2, then a factor K is incorporated in the formula. Calculations involving this case are not common in the API 510 exam.

The formulae used for the required minimum thickness of ellipsoidal heads in UG-32 (d) are:

(Semi)Ellipsoidal head

ASME VIII section UG-32 (d)

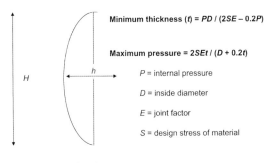

Minimum thickness $(t) = PD / (2SE - 0.2P)$

Maximum pressure $= 2SEt / (D + 0.2t)$

P = internal pressure

D = inside diameter

E = joint factor

S = design stress of material

In a 'standard' 2:1 ellipsoidal head:
Major axis $(H) = 2 \times$ minor axis (h)

For a non-standard ellipsoidal head a K-factor is involved. See API 510 (7.4.6.3)

Figure 9.6 Ellipsoidal head geometry

Minimum thickness $(t) = PD / (2SE - 0.2P)$
Maximum pressure $= 2SEt / (D + 0.2t)$

where

P = internal pressure
D = inside diameter (notice how the diameter replaces the R_i used for shells)
E = joint efficiency
S = allowable stress of the material

Ellipsoidal head calculation example

Here is an example for a 2:1 ellipsoidal head, using similar figures from the previous example. Guides:

D = inside diameter of 60 in
P = pressure of 250 psi (MAWP)
E = 0.85 (double-sided butt weld with spot examination (UW-12))
S = 15800 psi

What thickness is required to resist the internal pressure?

Using $t = PD/(2SE - 0.2P)$
Thickness $(t) = 250 \times 60 / [(2 \times 15800 \times 0.85) - (0.2 \times 250)]$

t = 0.56 in   ANSWER

Again, we can work the equation from the other perspective, where you are given a material thickness and you have to determine the MAWP.

Assuming a given head thickness of 0.625 in
Using maximum pressure (MAWP) $= 2SEt / (D + 0.2t)$

Pressure $= P = 2 \times 15800 \times 0.85 \times 0.625 / [60 + (0.2 \times 0.625)]$

P = 279 psi   ANSWER

9.5.3 Torispherical heads UG-32 (e)

Note the two restrictions on physical dimension of the head, given in UG-32(e) (see Fig. 9.7):

- Knuckle radius = 6 % of the inside crown radius
- Crown radius = outside diameter of skirt

These are long-standing restrictions based on well-established design principles for torispherical heads. The formulae for the required minimum thickness and MAWP given in UG-32 (e) are:

Thickness $t = 0.885PL/(SE - 0.1P)$
Pressure $P = SEt/(0.885L + 0.1t)$

where

L = inside spherical radius of the flatter part of the head (called the crown radius).

Torispherical head example
Given:

L = inside spherical (crown) radius of 30 in
P = pressure of 250 psi (MAWP)
$E = 0.85$
$S = 15\,800$ psi

$$\text{Thickness required } (t) = \frac{0.885 \times 250 \times 30}{(15\,800 \times 0.85) - (0.1 \times 250)}$$

$t = 0.495$ in   ANSWER

Alternatively, to find P using a given head thickness of 0.625 in:

$$\text{Pressure } (P) = \frac{15\,800 \times 0.85 \times 0.625}{(0.885 \times 300) + (0.1 \times 0.625)}$$

$P = 315$ psi   ANSWER

Torispherical head

ASME VIII section UG-32 (e)

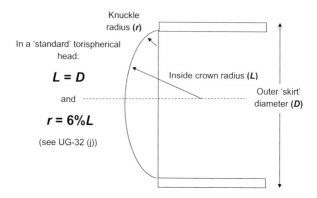

Thickness = $t = 0.885PL / (SE - 0.1P)$

Pressure = $P = SEt / (0.885L + 0.1t)$

Figure 9.7 Torispherical head geometry

9.5.4 Hemispherical heads UG-32 (f)

Limitations apply of thickness and pressure as before. The formulae for the required minimum thickness are given in UG-32 (f) (see Fig. 9.8):

$$t = PL/2SE - 0.2P$$
$$P = 2SEt/(L + 0.2t)$$

This time, L is the spherical inside radius (note that there is no crown or knuckle radius as the head is hemispherical; i.e. a half circle).

Hemispherical head example
Given:

Internal pressure $(P) = 200$ psi

ASME VIII Pressure Design

Hemispherical head

ASME VIII section UG-32 (f)

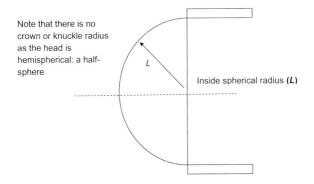

Thickness = $t = PL / (2SE - 0.2P)$

Pressure = $P = 2SEt / (L + 0.2t)$

Figure 9.8 Hemispherical head geometry

Allowable stress (S) = 15 000 psi
Spherical radius (L) = 60 in
Joint efficiency (E) = 1.0

Required thickness (t) = 200 × 60 / [(2 × 15 000 × 1)
− (0.2 × 200)]

t = **0.4 in** **ANSWER**

Alternatively, calculating the maximum allowable pressure for a given thickness of, say, 0.5 in:

Pressure (P) = 2 × 15 000 × 1 × 0.5/60 + (0.2 × 0.5)

P = **250 psi** **ANSWER**

9.5.5 Conical heads (without knuckle) UG-32 (g)

Only conical heads without a moulded knuckle are covered in the API 510 syllabus. A cone with a knuckle is known as a toriconical head. They are covered in UG-32 (g) but are not in the API 510 syllabus. In practice, conical heads of any type rarely appear as exam questions, but here they are for reference.

The formulae for the required minimum thickness given in UG-32 (g) are:

$t = PD/2 \cos \alpha (SE - 0.6P)$
$P = 2SEt \cos \alpha / [(D + 1.2 t \cos \alpha)]$

where alpha (α) is the half-cone angle of the cone.

Example of conical head calculation

Given:
  Internal pressure (P) = 300 psi
  Inside diameter of cone (D) = 40 in
  Allowable stress (S) = 12 000 psi
  Joint efficiency (E) = 0.85
  Cone half angle (α) = 30°
  Cosine of 30° = 0.866

Calculating required thickness (t):

$t = PD/2 \cos \alpha (SE - 0.6P)$

Thickness (t) = 300 × 40/[2 × 0.866 × (12 000 × 0.85 − (0.6 × 300)]

$t = 0.69$ in   ANSWER

Alternatively calculating the maximum allowable pressure for a given head thickness of, say, 0.75 in:

$P = 2SEt \cos \alpha / [(D + 1.2 t \cos \alpha)]$

Pressure (P) = 2 × 12000 × 0.85 × 0.75 × 0.866/ [40 + (1.2 × 0.75 × 0.866)]

$P = 325$ psi   ANSWER

ASME VIII Pressure Design

In recent years, conical head questions have appeared rarely, if ever, in the API 510 exam.

9.5.6 Corrosion allowance

Remember that none of the thicknesses calculated in the above calculations have included any allowance for the corrosion of the material. This must be added to the required thickness obtained from the calculation. Normally it is left to the particular owner/user to set his or her own requirements. This can vary greatly, depending on the type of material, from 0.25 in for carbon and low alloy steels to no corrosion allowance at all for stainless steel type materials.

The corrosion allowance is important when calculating the remaining life of a vessel under API 510 guidelines (as we will see in later calculations).

Now try these familiarization questions on shell and head calculations.

9.6 Set 1: shells/heads under internal pressure familiarization questions

Q1. ASME VIII UG-27: shells under internal pressure

The stress trying to split a vessel shell longitudinal weld open is called?

(a) Hoop stress ☐
(b) Circumferential stress ☐
(c) Longitudinal stress ☐
(d) (a) and/or (b) above ☐

Q2. ASME VIII UG-27: shells under internal pressure

The stress formulae of UG-27 relevant to cylinders are used for?

(a) 'Thin' cylinders ☐
(b) 'Thick' cylinders ☐
(c) Seamed cylinders with longitudinal welds only ☐
(d) (a) and (b) above ☐

Q3. ASME VIII UG-27: shells under internal pressure

The main cylinder dimension used in the UG-27 cylinder formulae is?

(a) Internal diameter ☐

(b) External diameter ☐
(c) Internal radius ☐
(d) External radius ☐

Q4. ASME VIII UG-27: shells under internal pressure
The parameter E used in the UG-27 cylinder formulae is?

(a) Joint efficiency ☐
(b) Yield strength ☐
(c) Young's modulus of elasticity ☐
(d) Allowable stress ☐

Q5. ASME VIII UG-27: shells under internal pressure
A vessel with a longitudinally seamed shell circumferentially welded to hemispherical heads is pressurized internally until it fails. Which of these formulae would you use to calculate the pressure at which the split would occur?

(a) $P = 2SEt/(R + 0.2t)$ ☐
(b) $P = 2SEt/(R - 0.4t)$ ☐
(c) $P = SEt/(R + 0.6t)$ ☐
(d) Either (b) or (c) above ☐

Q6. ASME VIII UG-27: shells under internal pressure
Which of these vessels could you *not* use the UG-27 formulae for (using a given joint efficiency of $E = 1$) if the material has an allowable stress at its design temperature of 20 ksi?

(a) $P = 0.5$ psi ☐
(b) $P = 100$ psi ☐
(c) $P = 2000$ psi ☐
(d) $P = 8000$ psi ☐

Q7. ASME VIII UG-27: shells under internal pressure
A vessel has the following given parameters

OD = 36 in
Wall thickness = 1 in
Joint efficiency $E = 1$
Allowable stress at design temperature = 20 ksi

What is the maximum allowable working pressure (MAWP) at its design temperature?

(a) 546 psi ☐
(b) 895 psi ☐
(c) 1136 psi ☐

ASME VIII Pressure Design

(d) 2409 psi ☐

Q8. ASME VIII UG-32 and UG-16 (b): heads under internal pressure

What is the absolute minimum allowable thickness of a vessel head of any design, irrespective of service fluid and material, if it is to have a corrosion allowance of 2 mm?

(a) 3.5 mm ☐
(b) 4 mm ☐
(c) 6 mm ☐
(d) 8 mm ☐

Q9. ASME VIII UG-32: heads under internal pressure

Which of these head designs has a knuckle radius and crown radius?

(a) Hemispherical ☐
(b) Any ellipsoidal design ☐
(c) Torispherical ☐
(d) Ellipsoidal with a t/L ratio of not less than 0.002 ☐

Q10. ASME VIII UG-32 (d): ellipsoidal head design

What is the minimum required thickness (t_{corroded}) of a 2:1 ellipsoidal head of $t/L > 0.002$ with the following dimensions:

D = inside diameter of 40 inches
P = pressure of 300 psi
E = 0.85 (double-sided butt weld with spot examination (UW-12))
S = 15 800 psi

(a) 0.4 in ☐
(b) 0.45 in ☐
(c) 0.55 in ☐
(d) None of the above ☐

9.7 ASME VIII: MAWP and pressure testing

We will now look at another of the major topics of the UG section of ASME VIII, the determination of MAWP (maximum allowable working pressure). This fits together with internal and external pressure calculations and influences the related subject of pressure testing. MAWP calculations are mathematically straightforward but cover a few different interconnected areas such as:

- Weld joint efficiency
- Static pressure head
- Calculating the MAWP for a corroded vessel and/or
- Calculating the required minimum thickness of a vessel subject to a known internal pressure

The last two in the above list are simply different ways of looking at the same thing. Both use the concept of a weld joint efficiency and a method to take into account the effect of static head pressure in the vessel, as well as pressure imposed from an external source or process.

9.7.1 MAWP

MAWP is an acronym used only in ASME and API codes. It is comparable, but not identical, to the term 'design pressure' preferred by most other non-US codes (see Fig. 9.9, and Fig. 4.7 in Chapter 4). Think of this way of understanding the ASME view of it:

- *Design pressure* is a nominal value of pressure provided by (for example) a process engineer or contractor to a vessel designer. This pressure is the minimum required in order for the vessel to fulfil its process function.
- The vessel designers then respond by designing a vessel based around MAWP because this is the parameter referred by ASME VIII when calculating the required thickness (t_{min}) of the pressure envelope components.
- From the above you can see that MAWP must be equal *to or greater than* 'design pressure'. It cannot be lower or the vessel will not meet its design requirement. You can see this relationship between design pressure and MAWP set out in ASME VIII UG-98 and API 576.

ASME VIII Pressure Design

In US codes, **DESIGN PRESSURE** is an almost nominal value, chosen at the pre-design stage to 'describe the process function' requirements of a vessel before code calculations are carried out

When design calculations are done, the concept of design pressure is superseded in importance by **MAWP**, which is either the same or greater than design pressure (see API 576 section 3.3.3)

MAWP is always measured at the *top* of the vessel

(at its operating temperature condition and in its corroded state)

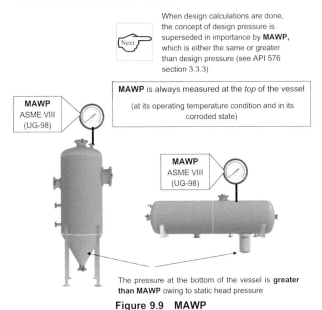

The pressure at the bottom of the vessel is **greater than MAWP** owing to static head pressure

Figure 9.9   MAWP

9.7.2   Where is MAWP measured?

By convention (and *only* by convention), MAWP is always measured at the top of a vessel. You can think of it as shown in Fig. 9.9 with the upper point of a vessel (vertical or horizontal) being fitted with a 'MAWP gauge'.

If the vessel is full of water the weight of the water (called the *static head*) causes the pressure at the bottom of the vessel to be greater than that at the top. Hence when designing the thickness of the bottom of the shell and the lower head the additional pressure due to the static head must be taken into account, over and above that pressure from the MAWP.

9.7.3 Pressure testing

The most common pressure test is the standard hydraulic test covered in ASME VIII UG-99 (see Fig. 9.10). Note how ASME use the term *hydrostatic* test – this term is strictly more applicable to atmospheric storage tanks, but that's what they use. Whereas, in earlier editions, ASME had used 1.5 × MAWP as the standard multiplier for test pressure this has now been amended to the following:

Test pressure (hydraulic) = 1.3 × design pressure × ratio of material stress values

$$\text{Ratio of material stress values} = \frac{\text{stress at test temperature (normally ambient)}}{\text{stress at design temperature}}$$

Remember that this test pressure is measured at the highest point of the vessel.

These stress values are taken from the tables of material stress values in ASME II(D) but are given in exam questions. Note that where a vessel is constructed of different materials that have different stress values, the *lowest* ratio of stress values is used.

This can be seen in an example:

Design pressure (MAWP) = 250 psi
Design temperature = 750°F
Material: carbon steel SA516-60

- Allowable stress value at room test temperature = 15 000 psi
- Allowable stress value at 750°F = 13 000 psi

Ratio of stress values = 15 000/13 000 = 1.154

Test pressure = 1.3 × 250 × 1.154
= 375 psi   ANSWER

The test medium for a hydrostatic test is normally water but other non-hazardous fluids may be used provided the test temperature is below its boiling point. Where combustible

ASME VIII Pressure Design

Test pressure (hydraulic) = 1.3 × MAWP × ratio of material stress values

Ratio of material stress values = $\dfrac{\text{Allowable stress at test temperature (ambient up to } 100°F)}{\text{Allowable stress at design (higher) temperature}}$

For API 510 exam purposes this ratio will always be **greater than 1**

ASME VIII (UG-99)

These stress values are normally given in the exam question. If a vessel is constructed of different materials that have different stress values, use the **lowest** given ratio

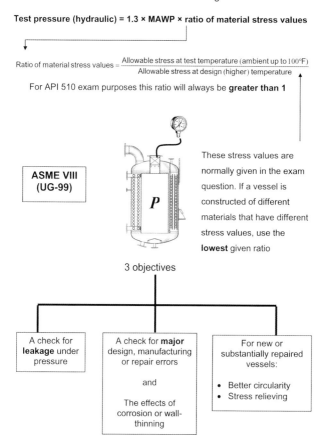

3 objectives

| A check for **leakage** under pressure | A check for **major** design, manufacturing or repair errors

and

The effects of corrosion or wall-thinning | For new or substantially repaired vessels:

• Better circularity
• Stress relieving |

Figure 9.10   The hydraulic (hydrostatic) test

fluids such as petroleum distillates are used with flash points of less than 110 °F (43 °C), these should only be used for tests carried out near atmospheric temperatures.

9.7.4 The hydrostatic test procedure

ASME VIII section UG-99 (g) gives requirements for the test procedure itself. This is a fertile area for closed-book examination questions. An important safety point is the requirement to fit vents at all high points to remove any air pockets. This avoids turning a hydrostatic test into a pneumatic test, with its dangers of stored energy.

Figure 9.11 shows the hydrostatic test procedure. A key point is that the visual inspection of the vessel under pressure is *not* carried out at the test pressure. It must be reduced back to MAWP (actually defined in UG-99 (g) as test pressure/1.3) before approaching the vessel for inspection. If it was a high-temperature test (> 120 °F), the temperature must also be allowed to reduce to this, before approaching the vessel.

Once the pressure has been reduced, all joints and connections should be visually inspected. Note how this may be waived provided:

- A leak test is carried out using a suitable gas.
- Agreement is reached between the inspector and manufacturer to carry out some other form of leak test.
- Welds that cannot be visually inspected on completion of the vessel were given visual examination prior to assembly (this may be the case with some kinds of internal welds).
- The contents of the vessel are not lethal.

In practice, use of these 'inspection waiver points' is not very common. Most vessels are tested and visually inspected fully as per the first sentences of UG-99 (g).

A footnote to UG-99 (h) suggests that a PRV set to 133 % test pressure is used to limit any unintentional overpressure due to temperature increases. Surprisingly, no PRV set to *test pressure* is required by the ASME code; you just have to be careful not to exceed the calculated test pressure.

ASME VIII Pressure Design

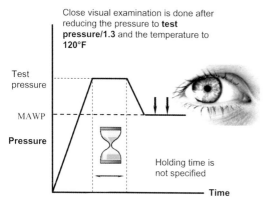

Temperature must be at least **30°F above MDMT** to avoid brittle fracture – ASME VIII (UG-99h). Note that API 510 overrides this – (API 510 (5.8.6.2) – and allows **10°F above MDMT** if the material is 2 in thick or less

Test gauges: some UG-102 guidelines

- Gauge full-scale deflection should be within 1.5 to 4 times the test pressure
- Gauges need to be calibrated
- One gauge is acceptable, as long as the operator can see it

---

Figure 9.11   The hydraulic test procedure

9.7.5 Pneumatic test

UG-100 covers the requirements for pneumatic testing. Look at the main points in Fig. 9.12. As a basic principle, ASME VIII does not recommend using pneumatic testing as an arbitrary alternative to a hydraulic test. The idea is that pneumatic testing is only used when it is absolutely necessary, when any of the following occurs:

- The vessel is not designed or supported in a way that it can be filled with water.
- A vessel cannot tolerate the presence of water.

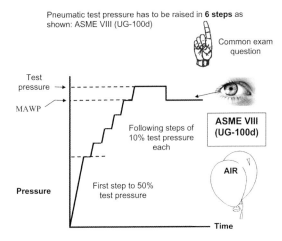

Test pressure (hydraulic) = 1.1 × MAWP × ratio of material stress values

Ratio of material stress values = $\dfrac{\text{Allowable stress at test temperature (ambient up to 100°F)}}{\text{Allowable stress at design (higher) temperature}}$

Temperature must be at least **30°F above MDMT** to avoid brittle fracture

Note that API 510 overrides this – (API 510 5.8.6.2) – and allows **10°F above MDMT** if the material is 2 in thick or less

Figure 9.12 The pneumatic test procedure

ASME VIII Pressure Design

- A previous hydraulic test has already been carried out (so the pneumatic test is only required as a type of leak test).

With pneumatic testing, the major consideration is safety. As air is compressible, a lot of energy will be stored in the vessel that can catastrophically release on failure of the vessel. Great care and hazard assessment is therefore needed before carrying out any pneumatic test.

To minimize any risk of brittle fracture, the test temperature should be at least 30 °F above the minimum design metal temperature (MDMT) and UG-100 gives a specific test procedure as follows.

The test pressure is *lower* than that for the hydro test:

Test pressure (pneumatic) = 1.1 × design pressure (MAWP) × ratio of material stresses

$$\text{Ratio of material stress values} = \frac{\text{allowable stress at test temperature (usually ambient)}}{\text{allowable stress at design temperature}}$$

As before, the stress values come from ASME II(D) so should be given in the exam question.

Section UG-100 (d) gives requirements for the pneumatic test procedure itself. This is a fertile area for closed-book examination questions. The steps are:

- Pressurize to 50 % test pressure.
- Increase in steps of approximately 10 % of test pressure until the required pressure is reached.
- Reduce to test pressure/1.1 and perform the visual inspection.

The only requirement for the test time is that the pressure must be held 'long enough to allow leakage to be detected'. In a similar way to hydraulic testing there are some 'unusual conditions' waivers allowed for the final visual inspection stage. It may be waived provided that:

- A leak test is carried out using a suitable gas.

Quick Guide to API 510

- Agreement is reached between the inspector and manufacturer to carry out some other form of leak test.
- All those welds that cannot be visually inspected on completion of the vessel are given visual examination prior to assembly.
- The vessel contents are not lethal.

YOU CAN IGNORE SECTION UG-101 'PROOFTESTING' AS IT IS NOT IN THE API 510 EXAMINATION SCOPE.

9.7.6 Test gauges UG-102

This is fairly straightforward and commonsense. Note that the requirements are based on a *single pressure gauge* to measure the test pressure. The requirements are:

- The gauge must be connected directly to the vessel (not via a network of pipes).
- It should be visible from where the pressure is applied (if it isn't, then an additional gauge must be provided).
- For large vessels, gauges that record the pressure measurements are recommended.
- The gauge should have an indicating range of not less than *$1\frac{1}{2}$ times the test pressure and not more than 4 times the test pressure*. This is to make sure that it gives an accurate reading.
- All gauges should be calibrated against either a dead weight tester or master gauge.

Now try these familiarization questions on MAWP and pressure testing.

9.8 Set 2: MAWP and pressure testing familiarization questions

Q1. ASME VIII UG-98: MAWP location

At what location is MAWP specified?

(a) Always at the top of a vessel ☐
(b) Always at the bottom of a vessel ☐
(c) At the location where the pressure gauge is located ☐

ASME VIII Pressure Design

(d) Any of the above, as long as you specify where it is ☐

Q2. ASME VIII UG-99: standard hydrostatic test
What is the current multiplier of hydrostatic test pressure compared to MAWP, excluding any temperature correction?

(a) 1.1 ☐
(b) 1.3 ☐
(c) 1.5 ☐
(d) It depends on the amount of RT that has been done (RT1, RT2, etc.) ☐

Q3. ASME VIII UG-99: standard hydrostatic test
A vessel with a MAWP of 1000 psi at a design temperature of 700 °F is hydrostatically tested at 1300 psi at ambient temperature. What is wrong with this?

(a) Nothing ☐
(b) The vessel has been overstressed ☐
(c) The vessel has not been stressed high enough to test its integrity ☐
(d) The low temperature will provide a brittle fracture risk ☐

Q4. ASME VIII UG-99: standard hydrostatic test
A vessel has a design pressure of 125 psi and a design temperature of 600 °F. The material allowable stresses (S) are shown in ASME II as follows:

- S at ambient temperature = 18.8 ksi
- S at 600 °F = 11.4 ksi

What is the correct hydrostatic test pressure at ambient temperature?

(a) 98.5 psi ☐
(b) 162.5 psi ☐
(c) 192 psi ☐
(d) 270 psi ☐

Q5. ASME VIII UG-99 (g): standard hydrostatic test
A vessel has a design pressure of 200 psi and a design temperature of 400 °F. The material allowable stresses (S) are shown in ASME II as follows:

- S at ambient temperature = 20 ksi
- S at 400 °F = 18 ksi

If it is hydrostatically tested at ambient temperature, at what pressure should an inspection be made of the weld joints and seams?

(a) 200 psi
(b) 222 psi
(c) 289 psi
(d) None of the above

Q6. ASME VIII UG-100 and UG-20: pneumatic test
Which of these statements is *false*?

(a) A pneumatic test places the vessel under less stress than a hydrostatic test
(b) A pneumatic test has more chance of resulting in brittle fracture
(c) A pneumatic test is more dangerous than a hydrostatic test
(d) A pneumatic test may be allowed on vessels that have not had 100 % RT

Q7. ASME VIII UG-100 (c): pneumatic test
A vessel is to be pneumatically tested to 100 psi with air. The vessel is manufactured from a material that has a minimum design metal temperature of 10 °F. What is the minimum temperature at which the vessel can be safely tested?

(a) 10 °F
(b) 30 °F
(c) 40 °F
(d) 68 °F

Q8. ASME VIII UG-100 (d): pneumatic test
What are the increments used to increase the pressure up to pneumatic test pressure?

(a) Increase gradually to 50 % design pressure followed by 10 % increments
(b) Increase gradually by 10 % increments
(c) Increase gradually to 50 % test pressure followed by 10 % increments
(d) Increase gradually to $1.3-x$ design pressure followed by 10 % increments

ASME VIII Pressure Design

Q9. ASME VIII UG-100 (d): pneumatic test

A vessel has a design pressure of 200 psi and a design temperature of 400 °F. The material allowable stresses (S) are shown in ASME II as follows:

- S at ambient temperature = 20 ksi
- S at 400 °F = 18 ksi

If it is pneumatically tested at ambient temperature, and as it is air and has no hydrostatic head effects, at what pressure should a leakage inspection be made of the weld joints and seams?

(a) 200 psi ☐
(b) 222 psi ☐
(c) 244 psi ☐
(d) 260 psi ☐

Q10. ASME VIII UG-102: test gauges

A vessel is to be pressure tested to 150 psi. Which of the following ranges should be used for the test gauge?

(a) 0–200 psi ☐
(b) 0–175 psi ☐
(c) 0–500 psi ☐
(d) 0–750 psi ☐

9.9 External pressure shell calculations

9.9.1 External pressure

So far, we have only looked at the design for vessels under internal pressure. We will now look at the topics of ASME VIII (section UG-28) covering vessels under *external pressure*. Many vessels are subject to vacuum conditions or have jackets, which can apply an external pressure to the shell (see Fig. 9.13). Details are given in UG-28 of ASME VIII for cylindrical vessels that may or may not have stiffening rings. Typical forms of cylindrical shells are shown in ASME VIII figure UG-28.1

External pressure calculations are *completely different* to those in UG-27 and UG-32 for internal pressure on heads and shells. This is because the mode of failure is completely different. Vessels under external pressure fail by *buckling*, a catastrophic (and fairly unpredictable) mechanism that is

Vessels subject to external pressure or internal vacuum fail by **buckling**. They use a completely different type of analysis to vessels under internal pressure

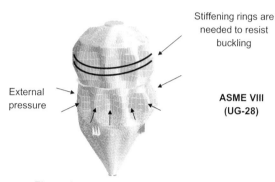

Figure 9.13 Vessel under external pressure

much more complex than the tensile stress failures that result from internal pressure.

Consequently, the concepts of allowable stress and joint efficiency do not play the same part that they do in internal pressure calculations on shells. Thankfully, the API 510 exam syllabus is limited to calculations relating to external pressure on cylindrical shells only. Heads are not covered (they are far too complicated).

Instead of simple stress equations formulae, external pressure calculations involve reading off charts to obtain two special factors A and B, and then using these factors in simple formulae to calculate either:

- the wall thickness (t) required to resist a given external pressure (P_a) or
- the maximum external pressure (P_a) that a vessel of given wall thickness (t) can resist.

9.9.2 External pressure exam questions

The API exam questions only require you to do a shortened method of assessment of external pressure conditions. This is because the two charts necessary to find the A and B calculation factors are not in ASME VIII (they are hidden away in ASME II, which is not part of the examination document package). In previous years the API 510 exam question paper occasionally contained extracts from these charts but more recently these have simply been replaced by given values of parameters A and B, making things much easier.

The easiest way to understand the UG-28 calculations themselves is to look at this worked example. Figure 9.14 shows the parameters for a vessel under external pressure operating at 300 °F:

- t = thickness of the shell = 0.25 in
- L = distance between stiffeners = 90 in
- D_o = shell outside diameter = 180 in

The first step is to calculate the values of the dimensional ratios (L/D_o) and (D_o/t):

$$L/D_o = 90/180 = \tfrac{1}{2}, \qquad D_o/t = 180 / 0.25 = 720$$

In a real design situation, these ratios would then be plotted on charts to give values of A and B. In this example, the charts would give values of A = 0.000 15 and B = 2250 (remember that you will generally be given these in an exam question).

From UG-28, the safe external pressure (P_a) is then calculated from the equation below:

$$P_a = 4B/3 \, (D_o/t) = 4 \times 2250/3 \times 720 = 4.2 \text{ psi}$$

Conclusion – the vessel is not suitable for full vacuum duty (−14.5 psi).

Quick Guide to API 510

EXTERNAL PRESSURE: WORKED EXAMPLE

Step 1. Assume the following parameters for a vessel under external pressure operating at 300 °F

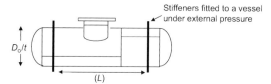

- t = thickness of the shell = 0.25 in
- L = distance between stiffeners = 90 in
- D_o = shell outside diameter = 180 in

The first step is to calculate the values of L/D_o and D_o/t

$L/D_o = 90/180 = \frac{1}{2}$ $D_o/t = 180/0.25 = 720$

These are plotted on the first chart to find the calculation factor **A**

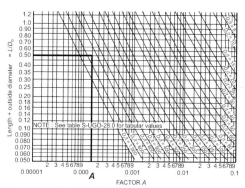

In this example the value of **A = 0.000 15**

Figure 9.14a   External pressure design *(continued over)*

ASME VIII Pressure Design

Next step is to plot this value of *A* on the second chart at the design temperature of the vessel, and read off the corresponding value of the second parameter **B**

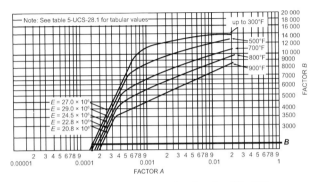

In this example taking 300 °F as the temperature, the value of **B = 2250**

Final step, use the ASME VIII (UG-28) formula to determine the maximum allowable external pressure P_a

$$P_a = 4B/3 \, (D_o/t) = 4 \times 2250 / 3 \times 720 = 4.2 \text{ psi}$$

Conclusion: the vessel is not suitable for full vacuum duty (−14.5 psi)

The calculation must be repeated either with an increased thickness or a lower value of *L*. (Add stiffening to the shell)

Figure 9.14b *(continued)* **External pressure design**

Quick Guide to API 510

Now try these familiarization questions.

9.10 Set 3: vessels under external pressure familiarization questions

Q1. ASME VIII UG-28: shells under external pressure

What is the relationship between the maximum internal pressure (P) a vessel can resist and the maximum external pressure (P_a) it can resist?

(a) $P_a = \frac{2}{3} P$ ☐
(b) $P_a = P \times$ factor A ☐
(c) $P = \frac{2}{3} P_a$ ☐
(d) There is no straightforward relationship ☐

Q2. ASME VIII UG-28 (b): lines of support

Where is the 'line of support' assumed to be in a vessel head with head depth h (where h excludes the straight flange part)?

(a) On the tan line (which is $h/2$ from the head-to-shell circ weld) ☐
(b) On the head-to-shell circ weld ☐
(c) $h/3$ into the head from the tan line ☐
(d) $h/3$ into the head from the head-to-shell circ weld ☐

Q3. ASME VIII UG-28(c): formula limits

Which formula should you use to determine the maximum allowable external pressure (P_a) for a cylinder with a wall thickness of 20 % of its external diameter D_o?

(a) $P_a = 4B/3(D_o/t)$ ☐
(b) $P_a = 2AE/3(D_o/t)$ ☐
(c) $P_a = 1.1/(D/t)^2$ ☐
(d) None of the above; it is not in the syllabus ☐

Q4. ASME VIII UG-28(c): formula limits

Limited data for a vessel are given as:

Outside diameter $D_o = 60$ in
Length between supports $L = 15$ feet
Factor $A = 0.000\,18$
Factor $B = 2500$

These are all the data you have. How thick does the vessel wall have to be to be suitable for use under full vacuum?

(a) $\frac{1}{8}$ in ☐

ASME VIII Pressure Design

(b) $\frac{1}{4}$ in ☐
(c) $\frac{3}{8}$ in ☐
(d) It is not suitable at all because of the large L/D_o ratio ☐

Q5. ASME VIII UG-28: *A*, *B* factors

In an exam question, from where are you likely to get the values for external pressure (P_a) factors A and B?

(a) You will be given them ☐
(b) From tables in your copy of ASME VIII Section UCS ☐
(c) From section 7 of API 510 ☐
(d) None of the above; you don't need them to calculate P_a ☐

9.11 Nozzle design

Two aspects of vessel nozzle design are included in the API 510 syllabus: nozzle compensation and weld sizing (see Fig. 9.15). By necessity the subjects covered are highly simplified in order to fit the required format of the exam questions. Over recent years the exam questions in this area have become both simpler and fewer in number, which is good news.

From a practical perspective, these subjects are in the API 510 syllabus owing to their relevance to vessel repairs. It is reasonable to expect that an inspector should be able to check design dimensions relating to a new or altered nozzle.

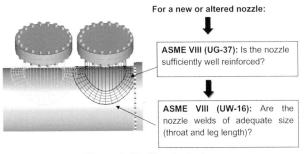

Figure 9.15 Nozzle design

9.11.1 Nozzle compensation

Exam questions on nozzle compensation centre around ASME VIII figure UG-37.1. In true ASME code style this figure incorporates a formidable amount of information on a single page. Note the figure itself. What it actually shows is a 'half-section diagram' – two halves separated by a vertical centreline (see Fig. 9.16). The left-hand side shows the configuration in which the nozzle is 'set through' the shell. You can ignore this as the set-through configuration is specifically excluded from the API syllabus. The right-hand side *is* in the syllabus – here the nozzle is *set on* to the shells, i.e. does not project through. The ASME VII term for this is *abuts*.

Now look at the equations and text below the diagram. This is (perhaps not so obviously) divided into two scenarios, this time separated by an imaginary horizontal line approximately halfway down the page. The top half covers the situation where the nozzle does not have a reinforcing element (compensation pad) and the lower half applies to when a reinforcing pad has been installed. Note some other points about this diagram.

- The area cut out to accommodate the nozzle is given the symbol A.
- The code uses the principle of the area replacement method (see Fig. 9.17). This is simple enough; the area cut out (A) must be replaced by metal available or added in other areas to restore the strength of the component. This replacement can be taken from a combination of four sources:
  - Excess material (above $t_{required}$) available in the shell (called A_1).
  - Excess material (above $t_{required}$) available in the nozzle (called A_2).
  - Material available in the welds. The number of welds obviously depends on whether or not there is a reinforcing pad fitted. These are called A_{41} and A_{42} (note that A_{43} as shown in ASME VIII figure UW-37.1

ASME VIII Pressure Design

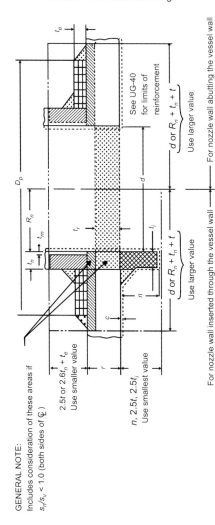

Figure 9.16a   ASME VIII nozzle diagram (Code figure UG 37.1). Courtesy of ASME *(continued over)*

Without Reinforcing Element

$A = d\, t_r\, F + 2 t_n\, t_r\, F(1 - f_{r1})$ — Area required

$A_1 = d(E_1 t - F t_r) - 2 t_n (E_1 t - F t_r)(1 - f_{r1})$
$A_1 = 2(t + t_n)(E_1 t - F t_r) - 2 t_n (E_1 t - F t_r)(1 - f_{r1})$ — Area available in shell: use larger value

$A_2 = 5(t_n - t_{rn})\, f_{r2}\, t$
$A_2 = 5(t_n - t_{rn})\, f_{r2}\, t_n$ — Area available in nozzle projecting outward; use smaller value

$A_3 = 5 t\, t_i\, f_{r2}$
$A_3 = 5 t_i\, t_i\, f_{r2}$
$A_3 = 2 h\, t_i\, f_{r2}$ — Area available in inward nozzle; use smallest value

A_{41} = outward nozzle weld = $(\text{leg})^2 f_{r2}$ — Area available in outward weld

A_{43} = inward nozzle weld = $(\text{leg})^2 f_{r2}$ — Area available in inward weld

If $A_1 + A_2 + A_3 + A_{41} + A_{43} > A$ — Opening is adequately reinforced

If $A_1 + A_2 + A_3 + A_{41} + A_{43} < A$ — Opening is not adequately reinforced so reinforcing elements must be added and/or thicknesses must be increased

Figure 9.16b ASME VIII nozzle diagram (Code figure UG 37.1). Courtesy of ASME (*continued over*)

ASME VIII Pressure Design

With Reinforcing Element Added

A = same as A, above — Area required

A_1 = same as A_1, above — Area available

$A_2 \begin{cases} = 5(t_n - t_m)f_{r2}t \\ = 2(t_n - t_m)(2.5t_n + t_e)f_{r2} \end{cases}$ — Area available in nozzle projecting outward; use smaller area

A_3 = same as A_3, above — Area available in inward nozzle

A_{41} = outward nozzle weld = $(\text{leg})2 f_{r3}$ — Area available in outward weld

A_{42} = outer element weld = $(\text{leg})2 f_{r4}$ — Area available in outer weld

A_{43} = inward nozzle weld = $(\text{leg})2 f_{r2}$ — Area available in inward weld

$A_5 = (D_p - d - 2t_n) t_e f_{r4}$ [Note (1)] — Area available in element

If $A_1 + A_2 + A_3 + A_{41} + A_{42} + A_{43} + A_5 > A$ — Opening is adequately reinforced

NOTE:
(1) This formula is applicable for a rectangular cross-sectional element that falls within the limits of reinforcement.

FIG. UG-37.1 NOMENCLATURE AND FORMULAS FOR REINFORCED OPENINGS
(This Figure Illustrates a Common Nozzle Configuration and Is Not Intended to Prohibit Other Configurations Permitted by the Code.)

Figure 9.16c *(continued)* ASME VIII nozzle diagram (Code figure UG 37.1). **Courtesy of ASME**

Quick Guide to API 510

The Area Replacement Method requires that:

Area removed (A) must be compensated for by the sum of

A1 +A2 + A4 +A5

i.e. You must have more spare area available than you have removed. If not, you add a reinforcing pad to make up the difference

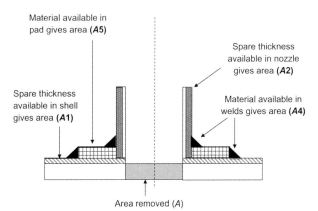

Note how this is a simplification of ASME VIII figure UG-37.1

Figure 9.17 The area compensation method

is not relevant as set-through nozzles are not in the syllabus).
- Material available (called A_5) in the reinforcing pad – if one is fitted.
- If the amount of compensation area available is more than that area removed (A), then the nozzle is adequately compensated, so no further compensation is required. This can be expressed in equation form, as shown near the bottom of figure UG-37.1.

An opening is adequately reinforced if:

$$A_1 + A_2 + A_3 + A_{41} + A_{42} + A_5 > A$$

ASME VIII Pressure Design

While not inherently difficult, a full compensation calculation involving all these variables is much too long for a 4-minute exam question, so actual exam questions tend to be heavily simplified, involving:

- Calculation of 'area required' A only.
- Calculations where most of the parameters are given, so only simple maths is required.
- Question about reinforcing limits.

9.11.2 Reinforcing limits

Reinforcing limits are the linear distances from a nozzle, beyond which adding reinforcement becomes ineffective. In practice, the size (diameter) of nozzle compensation pads is frequently chosen to coincide with the reinforcing limit, thereby achieving the most 'efficient' design.

Figure 9.18 shows the reinforcement limits, extracted from the code figure UG-37.1. Note how:

- There are two linear limits: one axially 'along' the vessel and the other extending radially outwards 'up the nozzle'.
- Both limits have two options for their calculated value. The axial limit uses the *larger* of its two options while the radial limit uses the *smaller* of its two options. In practice, it is usually the first term of each option that 'governs', but in an exam situation it is best to check both just in case.

Quick Guide to API 510

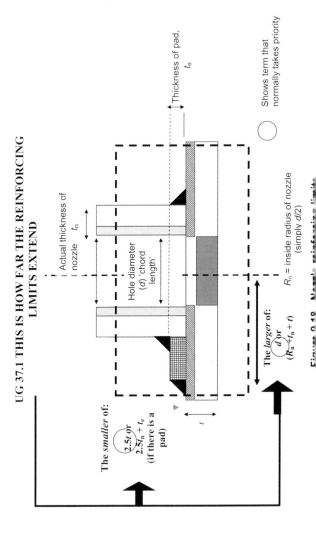

Figure 9.18 Nozzle reinforcing limits

ASME VIII Pressure Design

BACK IN THE DESIGN OFFICE 1922

Date: *June 1922 (the first of several repeat performances)*
The cast: *the Design Manager, the Draughtsman and a bank of silent assistants*
Surroundings: *The design office….an overbearing empire of faded mahogany wall panels, seasoned cherry-wood desks and the faint legacy of matured pipe tobacco*

Act One

'What units are we going to use for these new-fangled pressure vessel designs calculations?'

'It doesn't matter (rustling of papers and folding of diagrams)… maths is maths so it doesn't matter whether you use either USCS feet and pounds units or those SI units that probably haven't been invented yet, the answers will be the same' (prepares to leave) 'it's nearly four o'clock ... time I was in my garden.'

'No they won't, the main parameters can be aligned, sure; feet with metres, pounds with Newtons (and kilogrammes for good measure) but the accuracy of the calculation results will differ, depending upon the accuracy of the inputs.'

'Yes, that's right, 0.001 inch input accuracy can only give a maximum output accuracy of 0.001 inch (that's a *mil* to you) and similarly an input accuracy of 0.1 millimetre (mm) will give you a similar 0.1 mm accuracy of output, at best ... so that's fine. Anyway, about my new garden …'.

'The factor of four.'

'Four what?'

'Four times 1 mil gives 0.1 mm … they differ by a factor of four … 0.1 mm is four times bigger than one mil.'

'OK then, we'll use 0.01 mm instead.'

'Factor of two and a half, the other way … there are two and a half mils in 0.01 mm.'

'… and exactly what's wrong with that?'

'The measurement accuracy that you use affects the result of any real-world linear dimension of an irregular shape. Take your garden for instance … look at your own sketch; if we measure the boundary with a ruler 4 feet long we get a total boundary length of 3465 feet.'

'About twice the length of your average suburban plot ... look how it curves elegantly around the water feature.'

'Exactly, if we were to use a ruler reduced by a factor of four (the imaginatively named one-foot ruler), then because of the way it measures more accurately around all those undulations and curves, we get a vastly different result. Look! It is now nearer 5128 feet. That proves it... the accuracy of any output depends *absolutely* on the resolution accuracy of the input (the length of the ruler).

'It's the ...'.

'Same garden, different answer.'

'Fact or opinion?'

'Fact. Try it.'

'Will that mean vessel design inaccuracies?'

'Truly it will.'

'No need to worry ... the show will go on. Manual calculations will damp out all these variations; they will hide your factors of four away like they were never there. People will just round everything up like they've always done and we'll end up with a single indisputable answer.'

'Ah well, I suppose you're right; good job there's no accurate calculators around.'

'Phwaa ... saw one just the other day ... buzzing and whirring in the corner, flashing lights and tickertape all over the place. Twice the size of my desk it was ... they'll never catch on.'

Exit left

Chapter 10

ASME VIII Welding and NDE

10.1 Introduction

This chapter is to familiarize you with the general welding approach contained in sections UW of ASME VIII. API 510 is for in-service inspection of vessels and therefore most welding carried out will be *repair* welds, rather than welds carried out on new systems. API 510 also makes it a *mandatory* requirement to comply with the welding rules contained in ASME VIII.

Take a quick look at the scope of sections UW-1 through to UW-65 of the code (which are all of the UW sections). Not all the numbers run consecutively; some are missing. Note the following:

- Section UW is not just about welding; there are design and NDE-related subjects in there as well.
- Coverage of welding processes really only starts properly at section UW-27. Before that, the content is more about welded joints themselves, rather than the processes used to weld them.

10.2 Sections UW-1 to UW-5: about joint design

You can think of these sections as an introduction to section UW (don't ask where UW-4 has gone). Fundamentally, they are about joint design rather than welding techniques but you need them for background information. The best way to understand these is not to read them in the order presented in the code. Start with UW-3 (and its UW-3 figure) and then move to UW-2, which explains the *restrictions* placed on these joint categories by four special categories of vessel service.

10.2.1 UW-3: welded joint categories

The background to these joint categories is that the ASME design codes (unlike some other vessel codes) are built around the idea of a *joint efficiency factor* denoted by the symbol E. The factor E appears in the internal pressure equations and depends on:

- The method of welding
- The amount of NDE carried out on the weld

Various categories of joints are identified, which (as we will see later) are given different joint efficiencies.

ASME VIII pressure vessel welded joints are given a letter designation A, B, C or D depending on their location in the vessel. The designations are described in section UW-3 and illustrated on the page afterwards in figure UW-3. Note how this figure contains all the practical weld joint types that are found in standard types of pressure vessels. The most critical welds are those classified as category A, as these are the ones that require the most NDE. The content of ASME VIII figure UW-3 is shown in Fig 10.1. Note these points about it shown in the annotations:

Cat A includes all longitudinal welds and critical circumferential welds such as hemispherical head to shell welds.
Cat B includes most circumferential welded joints including formed heads (other than hemispherical) to main shells welds.
Cat C includes welded joints connecting:
- flanges to nozzles or shell components.
- one side plate to another in a flat-sided vessel.

Cat D includes welded joints connecting nozzles to shells, heads or flat-sided vessels.

Note one specific point identified in figure UW-3: a Cat B angled butt weld connecting a transition in diameter (i.e. tapered section) to a cylinder is included as a special requirement provided the angle (see figure UW-3) does not

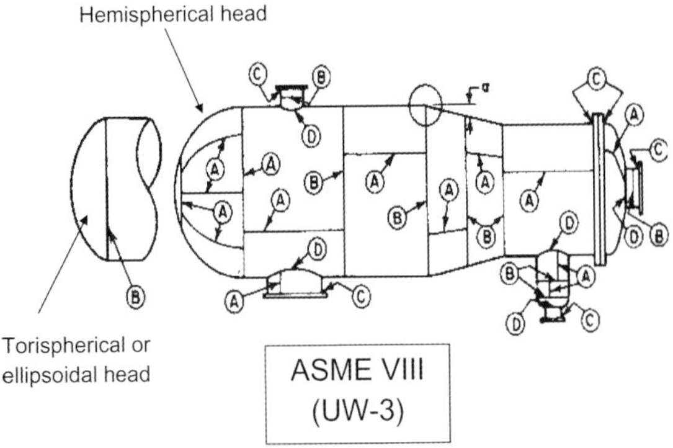

ASME VIII Welding and NDE

Figure 10.1 ASME VIII weld categories (courtesy ASME)

Note how the weld *Category* is governed by the location of the weld in the vessel. Don't confuse this with the weld *joint type* (described in figure UW-12)

Questions in the API 510 exam normally only involve categories A and B

exceed 30°. All the requirements of a butt-welded joint are applied to this angle joint.

This figure UW-3 is of limited use on its own. Its purpose is mainly to link with section UW-12: joint efficiencies. This section, with its accompanying table UW-12, allows you to determine the joint efficiency to use for a weld, as long as you know the category of weld (A–D), the weld joint arrangement (single or double groove, etc.) and the extent of NDE that has been carried out. We will see how to use this soon.

10.2.2 UW-2: service restrictions

Stepping back one section, UW-2 gives guidance on which types of pressure vessels/parts have restrictions on what type of weld should be used for each joint category. The four types of vessels referenced are:

(a) Vessels for lethal service (containing a lethal substance)
(b) Low-temperature vessels that require impact testing
(c) Unfired steam boilers

(d) Direct-fired vessels

These are referenced *a, b, c, d* in the code. Unfortunately, the page containing these is formatted using an impenetrable hierarchy of subheadings, making the explanations somewhat difficult to follow.

Most of the 'meat' in this section resides in the subsection covering the first category of vessel identified: (a) vessels for lethal service. The other three categories are more or less based on this first one, with some selected changes. Look at the requirement for lethal service and highlight the following key points:

- Butt-welded joints must be fully radiographed.
- Carbon or low alloy steel vessels need PWHT.
- Cat A joints need to be type 1(double vee or equivalent) welds (the types are given in table UW-12).
- Cat B joints can be either type 1 or type 2 (single vee with backing strip).
- Cat D must be full penetration welds.

There is a lot in this section about Cat C 'lap joint stub end' welds. These are an alternative to forged weld-neck flanges and not particularly common (even though there is a large section on them) except (presumably) on ASME VIII vessels for lethal fluid service.

Note how the (b) category, covering low-temperature vessels, has very similar requirements to the lethal service category, one slight difference being that Cat C welds have to be full penetration. The other two types of vessels, (c) unfired steam boilers and (d) direct-fired pressure vessels, are, again, similar, but with a few changes.

Remember the key point again:

- Most of the 'meat' in section UW-2 resides in subsection (a) *vessels for lethal service*. The other three categories are based on this first one, with some selected changes.

ASME VIII Welding and NDE

10.3 UW-12: joint efficiencies

UW-12 (text and table) are a core part of the welding requirements of ASME VIII. It is set out as shown in Fig. 10.2. Table UW-12 covers all types of gas and arc welding processes and spreads over two pages. Look at this table in your code and the notes 1–7 at the bottom of the second page of the table. These are often used as the subject matter for open-book examination questions in which table UW-12 is involved.

Now look at the body of table UW-12. It contains the following information:

- The weld type number – these range from types 1 to 8
- The joint description for each weld type
- Any limitations associated with a weld type
- The joint category (Cat A to D as already seen in UW-3)
- The degree of RT carried out, subdivided into three levels as follows:
  - Column (a) – full radiography

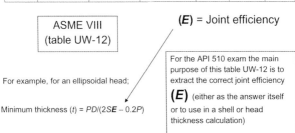

Figure 10.2 The format of ASME VIII table UW-12

- Column (b) – 'spot' radiography
- Column (c) – no radiography

These RT categories have their origin in section UG-116

Figure 10.2 shows where this is heading. Now look in more detail at ASME VIII table UW-12 and note the following key points:

- Look how joint efficiency E depends only on the type of joint and the extent of RT examination carried out on it. The three columns a, b and c cover the situation for conditions of full, 'spot' and no RT.
- The text is divided into six subsections (a to f) describing a variety of individual cases. You can think of this as elaborating on the content of table UW-12. Look through sections d, e and f in the code and note the points indicated below:

Item (d): seamless (forged) sections of vessels

This is a special technical case, but a common area for examination questions. Note what the code says:

(d) **Seamless vessel sections** are considered equivalent to welded vessel sections of the same geometry in which all Cat A welds are type 1. This applies to both head and shell sections.

Although (by definition) seamless shells or heads have no seams, there is still the need to decide a joint efficiency to use in the pressure design calculations. It is best not to think too deeply about this or you might think it doesn't quite make sense. ASME VIII obviously thinks it does. The E value to use is set out as follows:

- $E = 1.0$ when 'spot' RT has been done.
- $E = 0.85$ when 'spot' RT has not been done or when Cat A/B welds *joining seamless sections together* are of type 3, 4, 5 or 6.

ASME VIII Welding and NDE

Item (e): welded pipe or tubing

Following the same impeccable logic as for vessels, welded pipe or tubing is also treated in the same manner as seamless, but with the allowable tensile stress taken from the appropriate 'welded product' values in the material stress tables. The requirements of UW-12 (d) are applied as before.

10.4 UW-11: RT and UT examinations

Note the following principles of ASME VIII (UW-11).

10.4.1 RT levels

The three levels of RT are:

- Full RT (100 % of weld length of code identified welds)
- 'Spot' RT (a sample of weld length – minimum 6 inches)
- No RT (radiography not required at all)

The simple principle is that critical welds (those with a high risk of failure due to high stresses) will generally require full radiography to determine whether defects are present that could lead to failure. Welds that are less critical or less likely to fail if they contain a defect may not require full RT but will still require 'spot' RT. Joints that are not under internal pressure/high loads are less likely to fail and do not require any RT at all.

10.4.2 Minimum specified RT/UT requirements

This is the most important part of UW-11 (see Fig. 10.3). It gives the six situations where *full RT* is mandatory under UW-11 (a):

1. All butt welds in the shell and heads of vessels containing *lethal substances*.
2. All butt welds in vessels *over $1\frac{1}{2}$ in (38 mm) thick*, or exceeding the thicknesses prescribed in table UCS-57 (have a quick look forward to this table). Note the exemption from this: Cat B/C butt welds in nozzles and communicating chambers $\leq$ NPS 10 or $\leq 1\frac{1}{8}$ in (29 mm) wall thickness do not require RT. This is a GENERAL

Quick Guide to API 510

ASME VIII (UW-11)

UW-11 (a)(1): All shell/head butt welds for vessels in lethal service require full RT

UW-11 (a)(2): All butt welds above a certain thickness require mandatory full RT

.........and some other requirements of **UW-11 (a)** ...check your code and see what they are.....

The minimum thickness table is given in UCS-57 and relates to a material's P-number and group

Watch out for UW-11 (a)(5)(b) and its partner UW-12 (d)

Figure 10.3 Important principles of ASME VIII (UW-11)

EXEMPTION FOR SMALL NOZZLES. THESE REQUIRE NO RT AT ALL (so effectively escape 'under the radar' from the RT requirements of the various categories RT1, RT2, etc.).

3. All butt welds in the shell/heads of *unfired steam boilers* exceeding 50 psi (345 kPa).
4. This one covers *nozzles*. Full RT is required for all butt welds in nozzles, etc., attached to vessel sections or heads that require full RT under (1) or (3) above.

UW-11 (b): 'spot' RT
This says that you may use 'spot' RT (and use a lower joint efficiency E) instead of full RT on type 1 or 2 welds.

UW-11 (c): 'no' RT
As a principle, no RT is required when the vessel or vessel part is designed for external pressure only, or when the joint

ASME VIII Welding and NDE

design complies with UW-12 (c). Sections (d) to (f) cover a few additional UT requirements when specialist welding techniques are used (electrogas, electron beam, etc.).

Note this important point, hidden away at the end in (g):

(g) For RT and UT of butt welds, nominal thickness is the thickness of the thinner of the two parts to be joined. This nominal thickness may be needed to determine if RT or UT is required.

10.5 UW-9: design of welded joints

Finally, we will look at UW-9: design of welded joints. This is more a design issue than a welding one, and there is less to this section than first appears. The main content relates to two areas:

- Taper transitions between welded sections of unequal thickness
- 'Stagger' of longitudinal welds in vessels

Don't expect to have to consult detailed figures of weld joints in this section. There is only one figure UW-9, showing the requirement for tapers. Look first at this figure and notice the main points in its descriptive text UW-9 (c):

(c) **Tapered transitions** requires that tapered transitions must have a taper of at least 3:1 between sections if the sections differ by the smaller of:
- more than $\frac{1}{4}$ of the thickness of the thinner section or
- $\frac{1}{8}$ in (3.2 mm).

Now move to UW-9 (d). This requires that longitudinal joints between courses must be staggered by at least five times the thickness of the thicker plate unless 4 inches (100 mm) of the joints on either side of the circumferential joint is radiographed (probably unlikely). Note the requirement hidden in the body of the text referencing UW-42. It means that if the taper is formed by weld build-up the additional metal must be examined by PT/MT.

The two points above appear very frequently as API 510 exam questions. Now try these familiarization questions.

10.6 ASME VIII section UW-11 familiarization questions (set 1)

Q1. ASME VIII section UW-11 (a)
Which of the following may *not* require full RT?

(a) Some butt welds in the shell of vessels containing lethal substances ☐
(b) Some butt welds in the head of vessels containing lethal substances ☐
(c) Butt welds in the shell of unfired steam boilers with pressures > 50 psi ☐
(d) Category B or C butt welds in a non-lethal vessel ☐

Q2. ASME VIII section UW-11 (a)
'Full RT' under ASME VIII means a vessel must have:

(a) Radiography applied to all welds including fillets ☐
(b) 100 % of its welds radiographed ☐
(c) All of the welds required by code to be radiographed for their full length ☐
(d) All welds exceeding NPS 10 or $1\frac{1}{8}$ in (29 mm) radiographed ☐

Q3. ASME VIII section UW-11(a)
A vessel is manufactured from P4 Group 2 material. It has a shell thickness of 18 mm and is used to contain lethal substances. What RT is required for a shell-to-shell circumferential butt weld?

(a) Spot RT ☐
(b) No RT ☐
(c) Full RT ☐
(d) Any of the above can be used ☐

Q4. ASME VIII section UW-11 (a)
A vessel is manufactured from P1 Group 2 material. It has a shell thickness of $\frac{3}{4}$ in (19 mm) and does not contain a lethal substance. What RT is required for a shell longitudinal butt weld?

(a) Spot RT ☐
(b) No RT ☐

(c) Full RT ☐
(d) Any of the above can be used ☐

Q5. ASME VIII section UW-11 (a)(6) and UW-11 (d)

A 2 in (50 mm) thick P1 Group 1 material is welded using the Electrogas welding process to give a full-penetration groove weld in a vessel in non-lethal service. What are the NDE requirements?

(a) It must have full RT and then UT after PWHT ☐
(b) It must have full or spot RT ☐
(c) It must have spot RT and then UT after PWHT ☐
(d) Either (a) or (c) is acceptable ☐

10.7 Welding requirements of ASME VIII section UW-16

(a) UW-16 minimum requirements for attachment welds at openings

UW-16 deals with the configuration and size of vessel nozzles and attachments welded into vessels. It gives the location and minimum size of attachment welds and must be used in conjunction with the strength calculations required in UW-15. Note that weld strength calculations are *not* included in the API 510 syllabus. Note also that the terms *nozzles, necks, fittings, pads*, etc., mean almost the same thing. This is a fairly complex section about a fairly simple subject, and can be a bit tricky. Have a look at figure UW-16 spread over a few pages of ASME VIII. It produces a few open-book exam questions occasionally, but doesn't seem to be mainstream content.

(b) Symbols

This paragraph defines the symbols used in UW-16 and in figures UW-16.1 and UW-16.2. You will need to recognize these symbols in order to understand figure UW-16.1, the one with the most important content for exam purposes. The main ones are:

- t = nominal thickness of vessel shell or head
- t_n = nominal thickness of nozzle wall

- t_e = thickness of reinforcing plate
- t_w = dimension of attachment welds (fillet, single-bevel or single-J), measured as shown in figure UW-16.1
- t_{min} = the smaller of $\frac{3}{4}$ in (19 mm) or the thickness of the thinner of the parts joined by a fillet, single-bevel or single-J weld
- t_c = not less than the smaller of $\frac{1}{4}$ in (6 mm) or $0.7\, t_{min}$
- t_1 or t_2 = not less than the smaller of $\frac{1}{4}$ in (6 mm) or $0.7 t_{min}$

Don't be put off by these definitions, which look a little complicated. If you have difficulty differentiating between t_c, t_1, t_2 and t_w, review them in conjunction with figure UW-16.1 itself.

Paragraphs (c) and (d) cover the two main options for connecting nozzles to shells using Cat D welds.

(c) Necks attached by a full penetration weld

Paragraph (c) basically tells us that:

- A set-*on* nozzle will have full penetration through the *nozzle* wall.
- A set-*in* nozzle will have penetration through the *vessel* wall.

Examples of each are then given in sketches UW-16 (a) to UW-16 (e).

To ensure complete weld penetration, backing strips or similar must be used when welding from one side without any method of inspecting the internal root surface.

A nozzle requires a hole to be cut in the shell producing a weakened area that may require strengthening. This strengthening can be added in the following ways:

1. **By integral reinforcement (also known as self-reinforcement).** This consists of using a thicker shell and/or nozzle, forged inserts or weld build-up, which is integral to the shell or nozzle. Figure UW-16.1 sketches (a), (b), (c), (d), (e), (f-1), (f-2), (f-3), (f-4), (g), (x-1), (y-1) and (z-1) show examples.

ASME VIII Welding and NDE

2. **By adding separate reinforcement pads.** (These are also termed *compensation* pads). They are welded to the outside and/or inside surface of the shell wall to increase the thickness in the weakened area. Figure UW-16.1 sketches (a-1), (a-2) and (a-3) give examples of compensated nozzles.

Note the different ways of welding the reinforcement pads to the shell:

At the outer edge of the pad by a fillet weld, and either:

- where it meets a set-on nozzle, by a full penetration butt weld plus a fillet weld with minimum throat dimension $t_w \geqslant = 0.7 t_{min}$ or
- where it meets a set-in nozzle, by a fillet weld with minimum throat dimension $t_w \geqslant = 0.7 t_{min}$ (figure UW-16.1 sketch (h)).

At the outer and inner edge of the pad by a fillet weld if it does not meet the nozzle. The fillet weld will have a minimum throat dimension of $\frac{1}{2} t_{min}$. See figure UW-16.1 sketch (a-2) for an example of a fillet welded attachment.

Now try these familiarization questions.

10.8 ASME VIII section UW-16 familiarization questions (set 2)

Q1. ASME VIII section UW-16 (c) and sketches (a), (b). Necks attached by a full penetration weld

A nozzle is fitted abutting (i.e. *set-on*) the vessel wall. What is an acceptable method of attaching it?

(a) With a full penetration groove weld through the nozzle wall ☐
(b) With a full penetration groove weld through the vessel wall ☐
(c) With a partial penetration groove weld through the nozzle wall ☐
(d) Both (a) and (c) are acceptable ☐

Quick Guide to API 510

Q2. ASME VIII section UW-16 (c)
By what means can reinforcement be added to an opening in a pressure vessel?

(a) By integral reinforcement such as forged inserts ☐
(b) By using separate plates (compensation pads) ☐
(c) By using thicker shell material ☐
(d) All of the above are valid methods ☐

Q3. ASME VIII section UW-16 (c) figure 16.1(a)
A vessel is manufactured from P1 Grade 2 material. It has a shell thickness of 18 mm and is used to contain lethal substances. A set-on (abutting) nozzle of 12 mm thickness is attached using a category D full penetration weld with reinforcing fillet. What is the minimum required throat thickness of the reinforcing fillet weld?

(a) 12 mm ☐
(b) 8.4 mm ☐
(c) 6 mm ☐
(d) 12.6 mm ☐

Q4. ASME VIII section UW-16 (c)(2)(c)
A vessel has a shell thickness of $\frac{3}{4}$ in (19 mm). A set-on (abutting) nozzle of $\frac{1}{2}$ in (13 mm) thickness is attached using a category D full penetration weld. A reinforcing plate of $\frac{1}{4}$ in (6 mm) is required. What welds will be required to attach the reinforcing plate to the nozzle?

(a) A full penetration weld plus a fillet with a 4.2 mm throat ☐
(b) A full penetration weld plus a fillet with a 6 mm throat ☐
(c) A fillet weld with a 3 mm throat ☐
(d) A full penetration weld plus a fillet with a 3 mm throat ☐

Q5. ASME VIII section UW-16 (d)(1)
A nozzle of NPS 10 (DN 250) is inserted through a vessel wall and protrudes into the vessel by an amount equal to the nozzle thickness. The nozzle thickness is one half of the shell thickness. Which of the following weld combinations are acceptable to attach the nozzle?

(a) Partial penetration groove or fillet weld on inside and outside face ☐

(b) Partial penetration groove with reinforcing fillet on outside face ☐
(c) Partial penetration groove with reinforcing fillet on inside face ☐
(d) Any of the above is acceptable ☐

10.9 RT requirements of ASME VIII sections UW-51 and UW-52

Remember API 510? We saw some very general requirements for the NDE of *repair* welds, using the same principles to those for welds carried out on new systems. This made it a *mandatory* requirement to comply with the welding rules contained in ASME VIII. We have also seen that ASME VIII contains various requirements for RT, spread around several sections of the code. These included the RT 'marking' categories of UW-11 (RT1, RT2, etc.) and the joint efficiencies that result from the choice of RT scope, set out in UW-12. This worked on the general principle of ASME VIII of being able to *choose* the RT category to follow (within limits), as long as you are happy to live with the joint efficiency that results.

We will now look at some further RT requirements of ASME VIII as set out in sections UW-51 and UW-52. As with all parts of the ASME code, you will find the inevitable cross-references to other code sections, but they are not as extensive here as in some other parts of the code. Have a look at Fig. 10.4; this shows a summary of the referenced sections relating to RT.

Before progressing further with UW-51/UW-52 bear in mind the existence of table UCS-57: *radiographic examination* (see Fig. 10.5). This table is very important as it gives the nominal wall thickness above which it is *mandatory* to fully RT butt-welded joints. The content of UW-51 and UW-52 must therefore be seen against the background of these mandatory requirements.

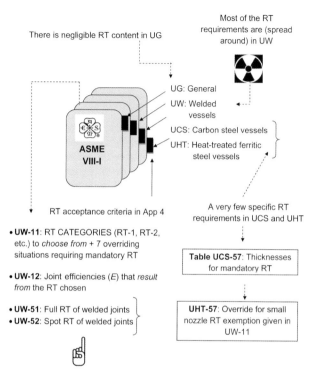

Figure 10.4 A summary of ASME VIII RT

10.9.1 UW-51: 'full' RT examination of welded joints

UW-51 and UW-52 are complementary sections. UW-51 deals with 'full RT' situations and UW-52 deals with those applicable to 'spot' RT.

Starting with UW-51 (a), this specifies that radiographed joints have to be examined in accordance with article 2 of section V. These are well defined and covered in the ASME V chapter of this book. There are a few differences that take

ASME VIII Welding and NDE

PRINCIPLE **UW-11 (a)(2)**: All butt welds above a certain thickness require mandatory full RT

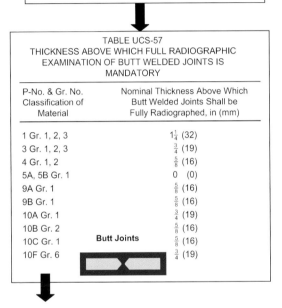

This minimum thickness table is given in **UCS-57** and relates to a material's P-number and group

TABLE UCS-57
THICKNESS ABOVE WHICH FULL RADIOGRAPHIC EXAMINATION OF BUTT WELDED JOINTS IS MANDATORY

P-No. & Gr. No. Classification of Material	Nominal Thickness Above Which Butt Welded Joints Shall be Fully Radiographed, in (mm)
1 Gr. 1, 2, 3	$1\frac{1}{4}$ (32)
3 Gr. 1, 2, 3	$\frac{3}{4}$ (19)
4 Gr. 1, 2	$\frac{5}{8}$ (16)
5A, 5B Gr. 1	0 (0)
9A Gr. 1	$\frac{5}{8}$ (16)
9B Gr. 1	$\frac{5}{8}$ (16)
10A Gr. 1	$\frac{3}{4}$ (19)
10B Gr. 2	$\frac{5}{8}$ (16)
10C Gr. 1	$\frac{5}{8}$ (16)
10F Gr. 6	$\frac{3}{4}$ (19)

Remember ....There is a general default level (for other materials) of $1\frac{1}{2}$ in (38 mm). This is shown in clause **UW-11 (a)(2)** itself

Figure 10.5   RT requirements of UCS-57. Courtesy of ASME

precedence over section V but they are mainly *procedural* (i.e. documentation/record-related). The main thrust of these is as follows:

- The manufacturer must retain a **complete set of radiographs** and records for each vessel until the Inspector has signed the Manufacturer's Data Report

- The manufacturer must certify that only **qualified and certified radiographers** and radiographic interpreters are used
- Radiographs will only be acceptable if the specified IQI hole or wire is visible

UW-51 (b) specifies conditions under which imperfections ('indications') are not acceptable and actions are to be taken. Note the following 'principle' point about the use of UT instead of RT:

Unacceptable imperfections must be repaired and re-radiographed. The Manufacturer can specify UT instead of RT, providing the original defect has been confirmed by UT to the satisfaction of the Authorized Inspector prior to making the repair. For material >1 in (25 mm) the User must also agree to its use. This UT examination must be noted under remarks on the Manufacturer's Data Report Form.

Note: Historically, the ASME code has been built on the premise of using RT as the main volumetric NDE method, but in recent years has started to accept UT as a viable alternative. In reality, however, RT still forms the basis of the ASME code's approach to integrity and it will probably take many years for this to change.

Defect acceptance criteria

ASME VIII, unlike some codes, *does* provide information on weld defect acceptance criteria. Note how these are slightly different for the 'full' and 'spot' RT scenarios. For 'full' RT, the following imperfections are *unacceptable*:

- Cracks, or incomplete fusion or penetration
- Elongated indications with lengths greater than the following
  - $\frac{1}{4}$ in (6 mm) for t up to $\frac{3}{4}$ in (19 mm)
  - $\frac{1}{3}t$ for t from $\frac{3}{4}$ in (19 mm to $2\frac{1}{4}$ in (57 mm)
  - $\frac{3}{4}$ in (19 mm) for t over $2\frac{1}{4}$ in (57 mm)

ASME VIII Welding and NDE

where

t = the thickness of the weld excluding reinforcement

For a butt weld joining two members having different thicknesses at the weld, t is the *thinner* of these two thicknesses. If a full penetration weld includes a fillet weld, the thickness of the throat of the fillet must also be included in t.

This section also concentrates on acceptance criteria for *aligned* indications. The following conditions are cause for *rejection*:

- A group of aligned indications with an aggregate length **greater than t in a length of 12t** unless the distance between the successive imperfections exceeds 6L, where L is the length of the longest imperfection in the group.
- Rounded indications in excess of those given in ASME VIII appendix 4.

Paragraph (c) gives examination requirements for real-time radioscopic examination. This is not a mainstream NDE technique in the pressure vessel industry so is unlikely to feature in the examination. Ignore it.

10.9.2 UW-52: 'spot' RT of welded joints

UW-52 begins with a note explaining the benefits and shortcomings of spot radiography. It basically points out that spot RT is useful for monitoring weld quality but can miss areas with weld defects present. If a weld must not have any defects in it then 100 % RT must be carried out.

Figure 10.6 shows the minimum extent of spot RT as specified by UW-52 (b). Look at these examples on its interpretation (it is fairly straightforward once you've got the idea):

- **A single vessel with 55 ft of weld** will have **two spots** examined, one spot for the 50 ft and one spot for the remaining 5 ft.

Quick Guide to API 510

- **Two identical vessels have 20 ft of weld each**. This gives a total of 40 ft and therefore only **one spot** needs to be taken on one of the vessels.
- **Two identical vessels have 40 ft of weld each**. This gives a total of 80 ft and therefore two spots need to be taken (one for the 50 ft and one for the remaining 30 ft). In this case one spot would be taken on each vessel.
- **Three identical vessels have 15 ft of weld each**. This gives a total of 45 ft and therefore only one spot needs to be taken on one of the vessels.

There are also some more general points on choosing the number and location of the spots.

UW-52 (c) gives acceptance criteria for spot RT. Note how they differ slightly from those in UW-52 for 'full' RT. The main points are as follows:

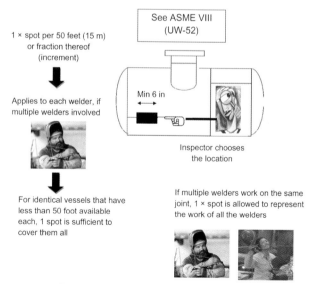

Figure 10.6 ASME VIII (UW-52) spot RT

ASME VIII Welding and NDE

- Cracks or zones of incomplete fusion or penetration are *unacceptable*.
- Slag inclusions or cavities with length $> 2/3t$ are *unacceptable* (the value of t is given in UW-52 (c)(2), which also gives limits on multiple inclusions or cavities $<2/3t$).
- Rounded indications *need not* be considered. (This is an important point ... these are only relevant when a weld needs full RT.)

Re-test of rejected welds

UW-52 (d) deals with *re-tests* when spot radiographs have failed their acceptance criteria. This uses the simple 2 for 1 principle, similar to that used in other codes (see Fig .10.7). Note the following main points in UW-52 (d):

1. When a spot radiograph is acceptable then **the entire weld increment** represented by it is acceptable.
2. When a spot radiograph shows a defect that is not acceptable then **two additional spots must be examined** in the same weld increment at locations chosen by the inspector:
    - If the two additional spots examined are acceptable the entire weld increment is acceptable provided the defect disclosed by the first radiograph is removed and the area repaired by welding. The weld-repaired area must then be radiographed again.
    - If either of the two additional spots examined are unacceptable then the entire increment of weld represented must *be rejected* and either:
        - replace the entire weld or
        - full RT the weld and correct any defects found.

Repair welding must be performed using a qualified procedure and in a manner acceptable to the Inspector. The re-welded joint, or the weld-repaired areas, must then be spot RT examined at one location in accordance with the foregoing requirements of UW-52.

Quick Guide to API 510

The 2 for 1 replacement rule

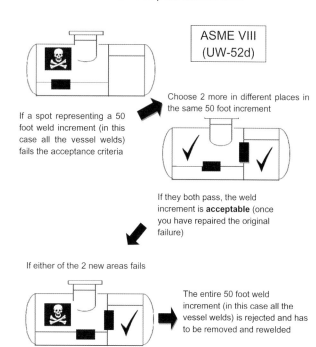

Figure 10.7 RT re-tests

10.10 ASME VIII section UW-51/52 familiarization questions (set 3)

Q1. ASME VIII section UW-51 (a)(4)
In ASME VIII Division 1 the final acceptance of the quality of a radiograph will be based on?

(a) The geometric unsharpness of the image ☐
(b) The ability to see the specified hole or designated wire of the IQI ☐
(c) The density ☐

ASME VIII Welding and NDE

(d) The film speed ☐

Q2. ASME VIII section UW-51 (a)(1)
When can a manufacturer destroy the radiographs relating to a vessel in lethal service?

(a) After 1 year ☐
(b) After 5 years ☐
(c) Once the inspector signs the Manufacturer's Data Report ☐
(d) Never, he must keep them with the technical file ☐

Q3. ASME VIII section UW-51 (b)(4)
Full RT has been carried out on a vessel marked as RT-1. A 1 in (25 mm) thick butt weld shows rounded indications. The required action will be to:

(a) Reject the weld ☐
(b) Accept the weld ☐
(c) Downgrade the vessel to RT-2 ☐
(d) Assess the rounded indications using the criteria in ASME VIII appendix 4 ☐

Q4. ASME VIII section UW-52 (c)
What is the minimum length of a spot radiograph?

(a) 3 in ☐
(b) 6 in ☐
(c) 9 in ☐
(d) 12 in ☐

Q5. ASME VIII section UW-52 (d)(2)
A spot radiograph fails the acceptance criteria. What should the API inspector request?

(a) An additional shot in the same location as the failure ☐
(b) Two additional shots in the same location as the failure ☐
(c) Two additional shots, one in the same location as the failure and the other remote from it ☐
(d) Two additional shots remote from the location of the failure ☐

Chapter 11

ASME VIII and API 510 Heat Treatment

Post-weld heat treatment (PWHT) is included in the API 510 examination syllabus, mainly in relation to vessel repairs. This makes sense as one of the roles of an API 510 inspector is to oversee repairs on behalf of the plant owner/user. Figure 11.1 shows the importance of this. Not all of the technical information relevant to PWHT is actually included in the API 510 document itself – most is contained in two well-defined (but separated) sections of ASME VIII.

This is not the end of the story. API 510 now adds (as it does with a few other topics) some requirements that can override the PWHT requirements of ASME VIII. The logic behind this is that whereas ASME VIII is a workshop-based construction code, under which PWHT can be done in a furnace under workshop conditions, API 510 deals with repairs, many of which will be carried out on site, where such closely controlled conditions are not possible. API 510 therefore provides easier alternatives that can be legitimately

PWHT refines the weldment grain structure after welding
- Reduces the chance of cracking
- Relieves residual stress

Figure 11.1 Post-weld heat treatment (PWHT)

ASME VIII and API 510 Heat Treatment

used during repairs. We will look at these parts of the content in turn.

11.1 ASME requirements for PWHT

Just about all of the main examination questions on PWHT that relate to ASME VIII are taken from either sections UCS-56 or UW-40. They are well separated in the code document but are cross-referenced directly to each other.

11.2 What is in UCS-56?

UCS-56 contains a couple of pages of text surrounding a group of seven or eight tables. Figure 11.2 shows a sample. The main content is in the tables; their purpose is to specify PWHT temperatures and holding times for different thicknesses of material. Each table covers different P-groups of material – because simplistically, the P-group is related to the tendency of a material to suffer post-weld cracking problems.

11.2.1 UCS-56 table notes

Don't ignore the half-page or so of notes underneath the various tables contained in UCS-56. They include information on either mandatory requirements or overriding exemptions, based mainly on material thickness.

Most exam questions (open book) will simply involve looking up the relevant PWHT time and temperature for a given material thickness in the correct 'P-group' table. Strictly, the material thickness to use is that of *nominal* thickness. This is defined not in UCS-56 but in UW-40 (f) – Fig. 11.2 shows the main points.

11.2.2 The UCS-56 text sections

There are a number of good open-book examination question subjects hidden away in the two pages of UCS-56 text. These relate to:

- The rate of heating of the PWHT furnace
- Allowable temperature variations in the furnace
- Furnace atmosphere

Quick Guide to API 510

This is a sample: UCS-56 contains multiple tables covering different material P-numbers

Material P-group	Minimum holding temperature °F	Minimum holding time for nominal thickness (see UW-40f)		
		Up to 2 in	Over 2 in to 5 in	Over 5 in
P1 Gr 1,2,3	1100	1 h/in (15 min minimum)	2 h plus 15 min for each additional inch over 2 in	2 h plus 15 min for each additional inch over 2 in

ASME VIII (UCS-56)

You need to know the governing (or 'nominal') thickness of the material to use when dealing with welded joints

For unequal-thickness butts, the nominal thickness is the *thinner*

Nominal thickness excludes cap reinforcement

Nominal thicknesses for PWHT purposes shown as (t)

ASME VIII (UW-40f)

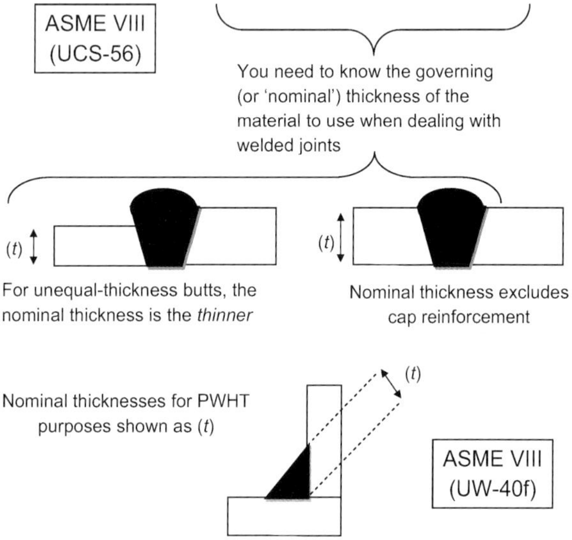

Figure 11.2 A specimen PWHT table UCS-56 and UW-40 nominal thickness

In addition to these clauses UCS-56 (f) and beyond gives six specific requirements relating to PWHT of weld repairs. In brief they are:

- The need for notification of repairs
- Maximum allowable depths of repair weld (38 mm for P1 Grades 1, 2, 3 and 16 mm for P3 Grades 1, 2, 3 materials)

- Excavation and PT/MT examination prior to repair
- Additional WPS requirements
- Pressure test after repair

Be careful not to misunderstand these requirements – ASME VIII is a construction code only, so the repairs it is referring to in UCS-56 (f) are repairs carried out as part of the *original manufacturing process*, *not* repairs carried out after in-service corrosion or some other damage mechanism. For in-service repairs ASME VIII requirements are overridden by the less stringent requirements of API 510 section 8, which does not place any limit on repair weld depth and divides repairs into temporary and permanent types.

11.2.3 The UW-40 text section
Whereas UCS-56 covers the times and temperature requirements for PWHT, UW-40 describes the procedures for *how* to do it. There are eight main options, some more practical than others.

11.3 API 510 PWHT overrides
API considers it a major advantage to be able to override the ASME VIII requirements for PWHT. Remember the logic behind this – API 510 relates to vessels once they are in use where the practicalities of site working probably will not allow manufacturing shop conditions to be reproduced so easily, if at all.

API 510 section 8.1.6.4 says that, in principle, repair welding must follow the requirements of ASME VIII (it means UW-40 and UCS-56) but opens the door to two overriding PWHT alternatives set out in API 510 section 8.1.6.4.2. This subsection has been progressively expanded and elaborated over recent code editions – you can see this in the out-of-balance subdivisions in the code clauses (it goes to a concentration-popping seven levels of subhierarchy, e.g. section 8.1.6.4.2.2.1).

The two methods of PWHT replacement (section 8.1.6.4.2) are:

- Replacement of PWHT by preheat
- Replacement of PWHT by controlled deposition (CD) welding methods

These are shown in Figs 11.3 to 11.5. Both work on the principle that the stress-relieving effects of PWHT can be achieved (albeit imperfectly) by providing the heat required in some other way than placing the repair in a furnace.

11.3.1 Replacement of PWHT by preheat

As the name suggests, this simply involves replacing PWHT with preheating the weld joint and then maintaining the temperature during the welding process. The maintained temperature serves to give sufficient grain refinement to reduce the chances of cracking when the weld is finished and allowed to cool down. While this technique provides sufficient grain refinement it is clearly not as good as full PWHT, so it is limited to materials of P1(Grade 1, 2, 3) and P3(Grade 1, 2) designations. These have a low risk of cracking anyway, owing to their low carbon content. P2 Grade 2 steels containing manganese and molybdenum are excluded, as they have a higher potential for cracking.

API exam questions normally centre around the parameters and restrictions of the preheat techniques. These are listed in API 510 section 8.1.6.4.2.2.1, and illustrated in Fig. 11.4.

11.3.2 Controlled deposition (CD) welding

This is sometimes known as *temper-bead welding* and is described in some detail in API 510 section 8.1.6.4.2.3. The principle is simple enough – when one layer of weld metal is laid down on top of another the heat from the upper one provides some heat treatment (grain refinement) to the weld underneath. A multilayer weld which is built up in this way will therefore be given an amount of grain refinement throughout its depth. The top layer of the final weld pass will not have anything above it to provide it with heat

ASME VIII and API 510 Heat Treatment

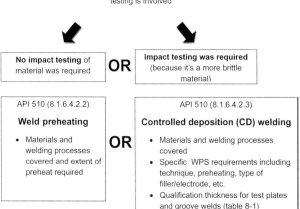

Figure 11.3 PWHT replacement options

treatment, so the solution is to grind it off. Figure 11.5 shows the idea.

The CD technique is considered to be a little better at replacing full PWHT than the preheat only alternative. It is therefore used for materials where the specification requires impact (notch toughness or Charpy) testing as a condition of their use in pressure equipment. The fact that impact tests were required indicates that the material has a tendency towards brittleness so the preheat method would not be good enough.

These two PWHT replacement techniques, preheat and CD welding, have become a mainstay of API codes. They are now mentioned in API 510, 570 and 653 and, we can assume, are commonly used in practice, although more commonly in the USA than elsewhere.

PWHT exam questions

PWHT replacement questions seem to be well-represented in the API exam question book. Questions on the validity of the two techniques, times, temperatures and heat-soak band dimensions crop up time and time again.

Now try these familiarization questions.

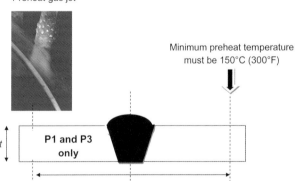

Figure 11.4 PWHT replacement by preheat

ASME VIII and API 510 Heat Treatment

> Override from API 510
> (8.1.6.4.2.3)

A tempering bead is placed on or partially overlapping the first bead ⇨ The heat from the temper bead heat treats the original HAZ and tempers the coarse grain structure, reducing the risk of cracking due to low toughness

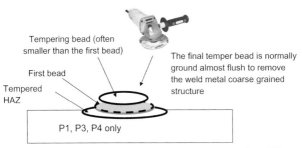

Figure 11.5 Controlled deposition (temper bead) welding

11.4 ASME VIII sections UCS-56 and UW-40: PWHT familiarization questions

Q1. ASME VIII section UCS-56
Under what circumstance can PWHT be omitted?

(a) When it is only a service requirement ☐
(b) When it is below the minimum thickness specified for the P-number in UCS-56 ☐
(c) When electron beam welding ferritic materials $> \frac{1}{8}$ in (3.2 mm) thick ☐
(d) When electroslag welding ferritic materials $> 1\frac{1}{2}$ in (38 mm) thick ☐

Q2. ASME VIII section UCS-56 (d)(2)
What is the maximum temperature *variation* permitted in a vessel during PWHT heating?

(a) 250 °F (139 °C) within any 15 ft (4.6 m) interval of length ☐
(b) 400 °F (204 °C) within any 15 ft (4.6 m) interval of length ☐
(c) 500 °F (260 °C) within any 15 ft (4.6 m) interval of length ☐
(d) 150 °F (67 °C) within any 15 ft (4.6 m) interval of length ☐

Q3. ASME VIII table UCS-56
A vessel is manufactured of P1 Group 2 material and is 4 in (102 mm) thick. What is the required PWHT holding temperature and minimum time at temperature?

(a) 1100 °F (593 °C) for 2 hours ☐
(b) 1100 °F (593 °C) for $2\frac{1}{2}$ hours ☐
(c) 1100 °F (593 °C) for 1 hour ☐
(d) None of the above ☐

Q4. ASME VIII table UCS-56
What is the PWHT minimum holding time for a 10 in (254 mm) thick P4 Group 2 material?

(a) 2 hours ☐
(b) 3 hours 15 minutes ☐
(c) 5 hours ☐
(d) 6 hours 15 minutes ☐

Q5. ASME VIII section UW-40 (f)(1) and (f)(5)(a)
What is used as the nominal thickness dimension of a full penetration butt weld joining a vessel head to a shell when the materials are of unequal thickness?

(a) The thickness of the thinnest part including the weld cap ☐
(b) The thickness of the thinnest part excluding the weld cap ☐
(c) The thickness of the thickest part including the weld cap ☐
(d) The thickness of the thickest part excluding the weld cap ☐

Chapter 12

Impact Testing

12.1 Avoiding brittle fracture

In any item of structural or pressure equipment there is a need to avoid the occurrence of *brittle fracture*. As we saw in API 571, brittle fracture is a catastrophic failure mechanism caused by the combination of low temperature and a material that has a low resistance to crack propagation at these temperatures. Under these conditions a material is described as having low toughness (or impact strength) – i.e. it is brittle. Impact strength is measured using a Charpy or Izod test in which a machined specimen is impacted by a swinging hammer. Figure 12.1 shows the situation.

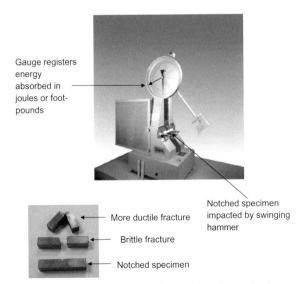

Figure 12.1 The Charpy impact toughness test

ASME design codes take a simplified pragmatic view of the avoidance of brittle fracture. Their view is that there are two levels to the checks required:

- First, there is a simple set of rules to determine if a material (and its design temperatures) actually needs impact testing or whether it can be assumed to be tough enough without being tested.
- Second, if it fails the first set of criteria and *does* need to be tested, what Charpy (Joules) or Izod (ft-lb) results need to be achieved for the material to be considered code-compliant.

Historically, you can expect a couple of questions in the API 510 exam relating to each of these criteria. The first criterion is a little more difficult to understand as there are several parts to it, and the code is not that easy to interpret on a casual reading. The second part is easier and just involves reading figures from tables, once you know where to find them. We will look at this now in UCS-66.

12.2 Impact exemption UCS-66

The main exam questions on this subject come from the tables and charts of UCS-66. Strictly, there are some opportunities for overall impact test exemptions that may apply before UCS-66 is even considered – these are tucked away in a totally separate part of the code: UG-20. Don't worry too much about these UG-20 requirements. They appear rarely, if at all, as exam questions, because they would divert attention away from UCS-66, which is where the impact strength questions usually come from.

In concept, UCS-66 is straightforward – the steps are as follows (see Fig. 12.2):

Step 1. For a given material determine, from figure UCS-66, whether it is covered by material curve A, B, C or D. Simply read this off the table, being careful to read the notes at the bottom of the table. In particular, notice that

Impact Testing

> Determine the material group A, B, C or D for any given material

GENERAL NOTES ON ASSIGNMENT OF MATERIALS TO CURVES:

Group A (a) Curve A applies to
  (1) all carbon and all low alloy steel plates, structural shapes, and bars not listed in Curves B, C, and D below;
  (2) SA-216 Grades WCB and WCC if normalized and tempered or water-quenched and tempered; SA-217 Grade WC6 if normalized and tempered or water-quenched and tempered.

Group B (b) Curve B applies to:
  (1) SA-216 Grade WCA if normalized and tempered or water-quenched and tempered
  SA-216 Grades WCB and WCC for thicknesses not exceeding 2 in. (51 mm), if produced to fine grain practice and water-quenched and tempered
  SA-217 Grade WC9 if normalized and tempered
  SA-285 Grades A and B
  SA-414 Grade A
  SA-515 Grade 60
  SA-516 Grades 65 and 70 if not normalized
  SA-612 if not normalized
  SA-662 if not normalized
  SA/EN 10028-2 P295GH as-rolled;
  (2) except for cast steels, all materials of Curve A if produced to fine grain practice and normalized which are not listed in Curves C and D below;
  (3) all pipe, fittings, forgings and tubing not listed for Curves C and D below;
  (4) parts permitted under UG-11 shall be included in Curve B even when fabricated from plate that otherwise would be assigned to a different curve.

Figure 12.2a   The UCS-66 steps. Courtesy of ASME
(continues on next page)

Quick Guide to API 510

Group C (c) Curve C
(1) SA-182 Grades 21 and 22 if normalized and tempered
SA-302 Grades C and D
SA-336 F21 and F22 if normalized and tempered
SA-387 Grades 21 and 22 if normalized and tempered
SA-516 Grades 55 and 60 if not normalized
SA-533 Grades B and C
SA-662 Grade A;
(2) all material of Curve B if produced to fine grain practice and normalized and not listed for Curve D below

Group D (d) Curve D
SA-203
SA-508 Grade 1
SA-516 if normalized
SA-524 Classes 1 and 2
SA-537 Grades Classes 1, 2 and 3
SA-612 if normalized
SA-612 if normalized
SA-738 Grade A
SA-738 Grade A with Cb and V deliberately added in accordance with the provisions of the material specification, not colder than −20°F (−29°C)
SA-738 Grade B not colder than −20°F (−29°C)
SA/AS 1548 Grades 7-430, 7-460, and 7-490 if normalized
SA/EN 10028-2 P295GH if normalized [see Note (g)(3)]
SA/EN 10028-3 P275NH

Now go to the relevant curve A, B, C, or D on the following graph ⟶

Figure 12.2a   The UCS-66 steps. Courtesy of ASME

Impact Testing

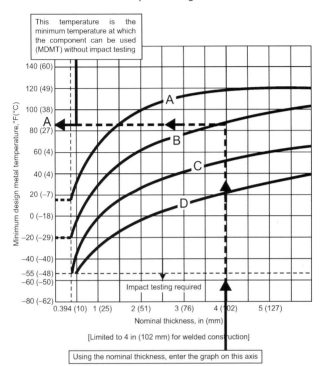

Figure 12.2b The UCS-66 steps. Courtesy of ASME

curve A provides a default for any relevant materials not listed in curves B, C or D. Note also how a material that has been normalized may be in a different group to the same material that is non-normalized. This is because normalizing affects the grain structure, and hence the brittle fracture properties.

Step 2. Determine the nominal thickness of the material. This is normally given in the exam question.

Step 3. In figure UCS-66 (for US units) or figure UCS-66M (for SI units), check the material thickness on the lower (horizontal) axis. Then read up the graph until you reach

Quick Guide to API 510

The low stress ratio temperature reduction

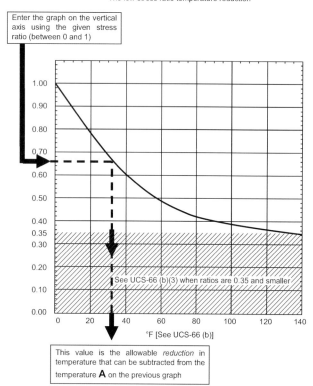

Figure 12.2c   The UCS-66 steps. Courtesy of ASME

the relevant curve A, B, C or D and read off the corresponding temperature on the vertical axis. Figure 12.2 (a) and (b) shows the procedure.

Step 4. Now the important part – the reading you just obtained on the vertical axis is the minimum temperature at which the component can be used (i.e. designed to be used) without requiring impact tests to check its resistance to brittle fracture. This design temperature is referred to as the minimum design metal temperature (MDMT) and is shown on the vessel nameplate.

If you are confused by this, just follow these two rules:

- If the required MDMT (i.e. the lowest temperature that you want the vessel to operate at) is *higher* than the temperature on the vertical axis of figure UCS-66, then impact tests are *not required* (because the material is not brittle at that temperature).
- Conversely, if the required MDMT is *lower* than the temperature on the vertical axis of figure UCS-66, then impact tests *are* required, to see if the material has sufficient toughness at that temperature.

Step 5. Check the figure UCS-66.1 'low stress ratio temperature reduction'. A feature of the ASME VIII-I code is that a material is considered less susceptible to brittle facture at a set temperature if the stress on the component is low. Technically, this is probably a disputable point, but the ASME codes have used it successfully for many years. The stress ratio is defined simply as the amount of stress a component is under compared to the allowable stress that the code allows for the material. It varies from 0 to 1.0, i.e. 0 % to 100 %, and in an exam question is normally given.

Figure UCS-66.1 and Fig. 12.2(c) show how the stress ratio reduction is used. This time, enter the graph on the vertical axis at the given stress ratio, move across to the curve and then read off the coincident temperature on the horizontal axis. This figure is the temperature *reduction* that can be subtracted from the previous temperature location on the vertical axis of figure UCS-66.

Step 6. Check the UCS-68 (c) 'voluntary heat treatment temperature reduction'. This is the final potential reduction allowed to the MDMT. Clause UCS-68 (c) (a few pages forward in the code) says that if a vessel is given voluntary heat treatment when it is specifically not required by the code (i.e. because the material is too thin or whatever), then a further 30 °F reduction may be applied to the MDMT temperature point identified on the original UCS-66 vertical axis. Note that this is *in addition* to any reduction available from the low-stress scenario.

Final step – general 'capping' conditions. Hidden in the body of the UCS-66 text are a couple of important 'capping' requirements. These occasionally arise in the API exam. The most important one is clause UCS-66 (b)(2). This is there to ensure that the allowable reductions to the impact exemption temperatures don't go too far. Effectively it 'caps' the exemption temperature at −55 °F for all materials. Note, however, the two fairly peripheral exceptions to this when the −55 °F cap can be overridden. These are:

- When the stress ratio is less than or equal to 0.35 (i.e. the shaded area of figure UCS-66.1). This is set out in UCS-66 (b)(3) and reinforces the ASME code view that components under low stress are unlikely to fail by brittle fracture.
- When the voluntary heat treatment of UCS-68 (c) has been done and the material is group P1.

The exam questions

Historically the API 510 exam questions on impact test exemption are pretty simple. They rarely stray outside the boundary of figure UCS-66 itself. The allowable reduction for low stress ratio and voluntary heat treatment are in the exam syllabus, but don't appear in the exam very often.

Now try these familiarization questions on impact test exemption.

12.3 ASME VIII section UCS-66: impact test exemption familiarization questions

Q1. ASME VIII section UCS-66
A stationary vessel is made from 3 in thick SA516 GR70 plate that has been normalized. The MDMT is 30 °F at 500 psig. Does this material require impact testing?

(a) No ☐
(b) Yes ☐
(c) Only if the vessel ID is less than 36 in ☐
(d) Only if the vessel ID is greater than 36 in ☐

Q2. ASME VIII section UCS-66
A vessel constructed of 'curve B' material is to be patch-plated with a fillet welded patch of the same material as the shell. The stress ratio is calculated as 0.64. The patch and vessel are 0.622 in thick with zero corrosion allowance. The MDMT is −15 °F. From the information given, does the repair require impact testing of the repair procedure?

(a) Yes ☐
(b) No ☐
(c) Yes, if optional PWHT is done ☐
(d) There is insufficient information in the question to decide ☐

Q3. ASME VIII section UCS-66
A 1.125 in thick lap-welded patch is of SA-515 Gr 70 P1 material. The vessel nameplate shows MDMT as 50 °F 'HT', denoting that the patch has been voluntarily heat-treated. The stress ratio is 1. From the information given, does the repair require impact testing of the repair procedure?

(a) Yes ☐
(b) No ☐
(c) Only if the optional PWHT is actually done ☐
(d) There is insufficient information in the question to decide ☐

Chapter 13

Introduction to Welding/API 577

13.1 Module introduction

The purpose of this chapter is to ensure you can recognize the main welding processes that may be specified by the welding documentation requirements of ASME IX. The API exam will include questions in which you have to assess a Weld Procedure Specification (WPS) and its corresponding Procedure Qualification Record (PQR). As the codes used for API certification are all American you need to get into the habit of using American terminology for the welding processes and the process parameters.

This module will also introduce you to the API RP 577 *Welding Inspection and Metallurgy* in your code document package. This document has only recently been added to the API examination syllabus. As a Recommended Practice (RP) document, it contains technical descriptions and instruction, rather than truly prescriptive requirements.

13.2 Welding processes

There are four main welding processes that you have to learn about:

- Shielded metal arc welding (SMAW)
- Gas tungsten arc welding (GTAW)
- Gas metal arc welding (GMAW)
- Submerged arc welding (SAW)

The process(es) that will form the basis of the WPS and PQR questions in the API exam will almost certainly be chosen from these.

The sample WPS and PQR forms given in the non-mandatory appendix B of ASME IX (the form layout is not strictly within the API 510 examination syllabus, but we will discuss it later) *only* contain the information for qualifying these processes.

13.2.1 Shielded metal arc (SMAW)

This is the most commonly used technique. There is a wide choice of electrodes, metal and fluxes, allowing application to different welding conditions. The gas shield is evolved from the flux, preventing oxidation of the molten metal pool (Fig. 13.1). An electric arc is then struck between a coated electrode and the workpiece. SMAW is a manual process as the electrode voltage and travel speed is controlled by the welder. It has a constant current characteristic.

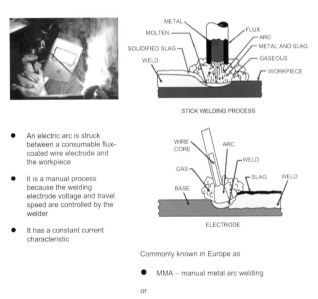

- An electric arc is struck between a consumable flux-coated wire electrode and the workpiece

- It is a manual process because the welding electrode voltage and travel speed are controlled by the welder

- It has a constant current characteristic

Commonly known in Europe as

- MMA – manual metal arc welding

or

- 'Stick' welding

Figure 13.1 The shielded metal arc welding (SMAW) process

13.2.2 Metal inert gas (GMAW)

In this process, electrode metal is fused directly into the molten pool. The electrode is therefore consumed, being fed from a motorized reel down the centre of the welding torch (Fig. 13.2). GMAW is know as a semi-automatic process as the welding electrode voltage is controlled by the machine.

Tungsten inert gas (GTAW)

This uses a similar inert gas shield to GMAW but the tungsten electrode is not consumed. Filler metal is provided from a separate rod fed automatically into the molten pool (Fig. 13.3). GTAW is another manual process as the welding electrode voltage and travel speed are controlled by the welder.

Submerged arc welding (SAW)

In SAW, instead of using shielding gas, the arc and weld zone are completely submerged under a blanket of granulated flux (Fig. 13.4). A continuous wire electrode is fed into the weld. This is a common process for welding structural carbon or carbon–manganese steelwork. It is usually automatic with

 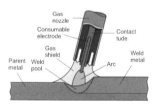

- An electric arc is struck between a continuously fed consumable solid electrode wire and the workpiece

- It is known as a semi-automatic process because the welding electrode voltage is controlled by the machine

Also known in Europe as

MIG – metal inert gas welding

or

MAG – metal active gas welding

Figure 13.2 The gas metal arc welding (GMAW) process

Introduction to Welding/API 577

 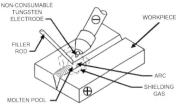

- An electric arc is struck between a non-consumable tungsten electrode and the workpiece. Filler rod is added separately

- It is a manual process when the welding electrode voltage and travel speed are controlled by the welder

Also known in Europe as

TIG – tungsten inert gas welding

or (rarely)

TAG – tungsten active gas welding

Figure 13.3   The gas tungsten arc welding (GTAW) process

the welding head being mounted on a traversing machine. Long continuous welds are possible with this technique.

Flux-cored arc welding (FCAW)
FCAW is similar to the GMAW process, but uses a continuous hollow electrode filled with flux, which produces the shielding gas (Fig. 13.5). The advantage of the technique is that it can be used for outdoor welding, as the gas shield is less susceptible to draughts.

13.3 Welding consumables

An important area of the main welding processes is that of weld *consumables*. We can break these down into the following three main areas:

- Filler (wires, rods, flux-coated electrodes)
- Flux (granular fluxes)
- Gas (shielding, trailing or backing)

There are always questions in the API examination about weld consumables.

Figures 13.6 to 13.11 show basic information about the main welding processes and their consumables.

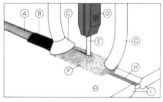
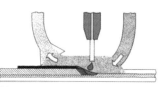

Schematic representation of submerged arc welding
A Finished weld
B Slag
C Powder removal
D Electrode holder
E Filler wire
F Flux
G Flux supply
H Root bead
I Parent metal

- An electric arc is struck between a reel-fed continuous consumable electrode wire and the work with the arc protected underneath a flux blanket

- It can be a semi-automatic, mechanized or automated process

Figure 13.4 The submerged arc welding (SAW) process

Introduction to Welding/API 577

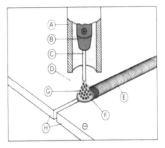

In FCAW the filler wire contains a flux. This protects the weld from the atmosphere by coating it in a slag (similar to SAW)

A Gas cup
B Electrode holder
C Filler wire
D Shielding gas
E Finished weld
F Weld pool
G Arc
H Parent metal

Figure 13.5 The flux cored arc welding (FCAW) process

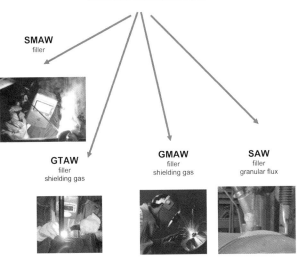

Figure 13.6 Welding consumables

Quick Guide to API 510

TYPES

- Basic: for low-hydrogen applications
- Rutile: for general purpose applications
- Cellulosic: for stovepipe (vertical down) applications

FILLER
Flux-coated electrodes

Figure 13.7 SMAW consumables

The American Welding Society (AWS) have a welding electrode identification system (see API 577 section 7.4 and appendix A). This is the system for SMAW electrodes

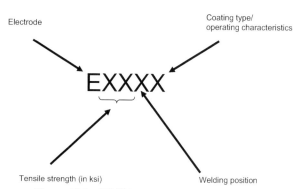

Figure 13.8 SMAW consumables identification

Introduction to Welding/API 577

FILLER:
Solid rods or wire

SHIELDING GAS:
Argon, helium or mixtures

Figure 13.9 GTAW consumables

FILLER:
Solid wire supplied on reels

SHIELDING GAS:
Inert gases – argon, helium, or mixtures. Active gases – carbon dioxide (CO_2) or Ar/CO_2 mixtures

Figure 13.10 GMAW consumables

FILLER:
Solid wire supplied on reels

FLUX:
Agglomerated: for low-hydrogen applications
Fused: for general applications

Figure 13.11   SAW consumables

Now try these two sets of familiarization questions about the welding processes and their consumables.

13.4   Welding process familiarization questions

Q1.   API 577 section 5.2
How is fusion obtained using the SMAW process?

(a) An arc is struck between a consumable flux-coated electrode and the work ☐
(b) An arc is struck between a non-consumable electrode and the work ☐
(c) The work is bombarded with a stream of electrons and protons ☐
(d) An arc is struck between a reel-fed flux-coated electrode and the work ☐

Q2.   API 577 section 5.1
Which of the following is not an arc welding process?

(a) SMAW ☐
(b) STAW ☐
(c) GMAW ☐
(d) GTAW ☐

Q3.   API 577 section 5.3
How is fusion obtained using the GTAW process?

(a) An arc struck between a consumable flux-coated electrode and the work ☐

Introduction to Welding/API 577

(b) An arc between a non-consumable tungsten electrode and the work ☐
(c) The work is bombarded with a stream of electrons and protons ☐
(d) An arc is struck between a reel-fed flux-coated electrode and the work ☐

Q4. API 577 section 5.3
How is the arc protected from contaminants in GTAW?

(a) By the use of a shielding gas ☐
(b) By the decomposition of a flux ☐
(c) The arc is covered beneath a fused or agglomerated flux blanket ☐
(d) All of the above methods can be used ☐

Q5. API 577 section 5.4
How is fusion obtained using the GMAW process?

(a) An arc struck between a consumable flux-coated electrode and the work ☐
(b) An arc between a non-consumable electrode and the work ☐
(c) The work is bombarded with a stream of electrons and protons ☐
(d) An arc is struck between a continuous consumable electrode and the work ☐

Q6. API 577 section 5.4
Which of the following are modes of metal transfer in GMAW?

(a) Globular transfer ☐
(b) Short-circuiting transfer ☐
(c) Spray transfer ☐
(d) All of the above ☐

Q7. API 577 section 5.6
How is the arc shielded in the SAW process?

(a) By an inert shielding gas ☐
(b) By an active shielding gas ☐
(c) It is underneath a blanket of granulated flux ☐
(d) The welding is carried out underwater ☐

Q8. API 577 section 5.6
SAW stands for:

(a) Shielded arc welding ☐
(b) Stud arc welding ☐
(c) Submerged arc welding ☐
(d) Standard arc welding ☐

Q9. API 577 sections 5.3 and 3.7
Which of the following processes can weld *autogenously*?

(a) SMAW ☐
(b) GTAW ☐
(c) GMAW ☐
(d) SAW ☐

Q10. API 577 section 5.3.1
Which of the following is a commonly accepted advantage of the GTAW process?

(a) It has a high deposition rate ☐
(b) It has the best control of the weld pool of any of the arc processes ☐
(c) It is less sensitive to wind and draughts than other processes ☐
(d) It is very tolerant of contaminants on the filler or base metal ☐

13.5 Welding consumables familiarization questions

Q1.
In a SMAW electrode classified as E7018 what does the 70 refer to?

(a) A tensile strength of 70 ksi ☐
(b) A yield strength of 70 ksi ☐
(c) A toughness of 70 J at 20 °C ☐
(d) None of the above ☐

Q2.
Which of the following does not produce a layer of slag on the weld metal?

(a) SMAW ☐
(b) GTAW ☐
(c) SAW ☐

Introduction to Welding/API 577

(d) FCAW ☐

Q3.
Which processes use a shielding gas?

(a) SMAW and SAW ☐
(b) GMAW and GTAW. ☐
(c) GMAW, SAW and GTAW ☐
(d) GTAW and SMAW ☐

Q4.
What type of flux is used to weld a low hydrogen application with SAW?

(a) Agglomerated ☐
(b) Fused ☐
(c) Rutile ☐
(d) Any of the above ☐

Q5.
What shielding gases can be used in GTAW?

(a) Argon ☐
(b) CO_2 ☐
(c) Argon/CO_2 mixtures ☐
(d) All of the above ☐

Q6.
Which process does not use bare wire electrodes?

(a) GTAW ☐
(b) SAW ☐
(c) GMAW ☐
(d) SMAW ☐

Q7.
Which type of SMAW electrode would be used for low hydrogen applications?

(a) Rutile ☐
(b) Cellulosic ☐
(c) Basic ☐
(d) Reduced hydrogen cellulosic ☐

Q8.
In an E7018 electrode, what does the 1 refer to?

(a) Type of flux coating ☐

(b) It can be used with AC or DC ☐
(c) The positional capability ☐
(d) It is for use with DC only ☐

Q9.

Which of the following processes requires filler rods to be added by hand?

(a) SMAW ☐
(b) GTAW ☐
(c) GMAW ☐
(d) SAW ☐

Q10.

Which of the following process(es) use filler supplied on a reel?

(a) GTAW ☐
(b) SAW ☐
(c) GMAW ☐
(d) Both (b) and (c) ☐

Chapter 14

Welding Qualifications and ASME IX

14.1 Module introduction

The purpose of this chapter is to familiarize you with the principles and requirements of welding qualification documentation. These are the Weld Procedure Specification (WPS), Procedure Qualification Record (PQR) and Welder Performance Qualification (WPQ). The secondary purpose is to define the essential, non-essential and supplementary essential variables used in qualifying WPSs.

ASME section IX is a part of the ASME Boiler Pressure Vessel code that contains the rules for qualifying welding procedures and welders. It is also used to qualify welders and procedures for welding to ASME VIII.

14.1.1 Weld procedure documentation: which code to follow?

API 510 (section 8.1.6.2.1) requires that repair organizations must use welders and welding procedures qualified to ASME IX and maintain records of the welding procedures and welder performance qualifications. ASME IX article II states that each Manufacturer and Contractor shall prepare written Welding Procedure Specifications (WPSs) and a Procedure Qualification Record (PQR), as defined in section QW-200.2.

14.2 Formulating the qualification requirements

The actions to be taken by the manufacturer to qualify a WPS and welder are done in the following order (see Fig. 14.1):

Quick Guide to API 510

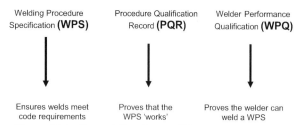

Figure 14.1 Formulating the qualification requirements

Step 1: qualify the WPS

- A preliminary WPS (this is an unsigned and unauthorized document) is prepared specifying the ranges of essential variables, supplementary variables (if required) and non-essential variables required for the welding process to be used.
- The required numbers of test coupons are welded and the ranges of essential variables used recorded on the PQR.
- Any required non-destructive testing and/or mechanical testing is carried out and the results recorded in the PQR.
- If all the above are satisfactory then the WPS is qualified using the documented information on the PQR as proof that the WPS works.

The WPS (see Fig. 14.2) is signed and authorized by the manufacturer for use in production.

Step 2: qualify the welder. The next step is to qualify the *welder* by having him weld a test coupon to a qualified WPS. The essential variables used, tests and results are noted and the ranges qualified on a Welder Performance Qualification (WPQ) (see Fig. 14.3).

Note that ASME IX does not require the use of preheat or PWHT on the welder test coupon. This is because it is the skill of the welder and his ability to follow a procedure that is being tested. The pre- and PWHT are not required because the mechanical properties of the joint have already been determined during qualification of the WPS.

Welding Qualifications and ASME IX

ASME IX QW 482 WPS format:

Company Name   MET Ltd   By: S. Hughes
Welding Procedure Specification No   SMAW-1   Date   01/04/2006   Supporting PQR No   SMAW-1
Revision No   0   Date   01/04/06

Welding Process(es)   SMAW   Type(s)   Manual
                                                   Automatic, Manual, machine or Semi Automatic

JOINTS (QW-402)

                                                                    Details
Joint Design   Single Vee Butt
Backing (Yes)   (No)   X   See production drawing
Backing Material (Type)
                  Refer to both backing and retainers

☐ Metal   ☐ Nonfusing Metal

☐ Nonmetallic   ☐ Other

Sketches, Production Drawings, Weld Symbols or Written Description should show the general arrangement of the parts to be welded. Where applicable, the root spacing and the details of weld groove may be specified. At the option of the manufacturer, sketches may be attached to illustrate joint design, weld layers and bead sequence, eg for notch toughness procedures, for multiple process procedures etc

BASE METALS (QW-403)

P-No.   1   Group No.   2   to P-No.   1   Group No.   2
OR
Specification type and grade
to
Specification type and grade
OR
Chemical Analysis and Mechanical properties
to
Chemical Analysis and Mechanical properties

Thickness Range:
   Base Metal: Groove   1/16" – 2"   Fillet
Pipe Diameter range: Groove   All   Fillet   All
Other

FILLER METALS (QW-404) Each base metal-filler metal combination should be recorded individually

Spec. No (SFA)   SFA 5.1
AWS No (Class)   E7016
F No   6
A-No   4
Size of filler metals   All
Weld Metal
Thickness range
       Groove   All
       Fillet   All
Electrode-Flux (Class) N/A
Flux Trade Name   N/A
Consumable Insert   N/A
Other

Figure 14.2a   WPS format

Quick Guide to API 510

ASME IX QW 482 (Back)

WPS No _SMAW-1___ Rev ___0__

POSITIONS (QW-405)

Position(s) of Groove _____

Welding Progression: Up _____ Down _____

Position(s) of Fillet _____

PREHEAT (QW-406)

Rate
Preheat Temp. Min _____
Interpass Temp. Max _____
Preheat Maintenance _____
(Continuous or special heating, where applicable, should be recorded)

POSTWELD HEAT TREATMENT (QW-407)

Temperature Range _____

Time Range _____

GAS (QW-408)
            Percent composition
Gas(es)   (Mixture)   Flow

Shielding _____

Trailing _____

Backing _____

ELECTRICAL CHARACTERISTICS (QW-409)
Current AC or DC _____   Polarity _____
Amps (Range) _____   Volts (Range) _____
(Amps and volts range should be recorded for each electrode size, position, and thickness, etc. this information may be listed in a tabular form similar to that shown below).

Tungsten Electrode Size and Type _____
                      (Pure Tungsten, 2% Thoriated, etc)
Mode of Metal Transfer for GMAW _____
                              (Spray arc, short circuiting arc, etc)
Electrode Wire feed speed range _____

TECHNIQUE (QW-410)
String or Weave Bead _____
Orifice or Gas Cup Size _____
Initial and Interpass Cleaning (Brushing, Grinding, etc) _____

Method of Back Gouging _____
Oscillation _____
Contact Tube to Work Distance _____
Multiple or Single Pass (per side) _____
Multiple or Single Electrodes _____
Travel Speed (Range) _____
Peening _____
Other _____

Weld Layer(s)	Process	Filler Metal		Current			Travel Speed Range	Other (remarks, comments, hot wire addition, technique, torch angle etc)
		Class	Diameter	Type Polarity	Amp Range	Volt range		

Figure 14.2b WPS format

Welding Qualifications and ASME IX

QW 483 PQR format

Company Name _____

Procedure Qualification Record No. _____ Date _____

WPS No _____

Welding Process(es) _____

Types (Manual, Automatic, Semi-Auto)

JOINTS (QW-402)

Groove Design of Test Coupon
(For combination qualifications, the deposited weld metal thickness that shall be recorded for each filler metal or process used)

BASE METALS (QW-403)
Material spec. _____
Type or Grade _____
P-No _____ to P-No _____
Thickness of test coupon: _____
Diameter of test coupon _____
Other _____

POSTWELD HEAT TREATMENT (QW-407)
Temperature _____
Time _____
Other _____

GAS (QW-408)
Percent composition
Gas(es)   (Mixture)   Flow Rate
Trailing ____ None ____
Backing ____ None ____
Backing ____ None ____

FILLER METALS (QW-404)
SFA Specification _____
AWS Classification _____
Filler metal F-No _____
Weld Metal Analysis A-No _____
Size of Filler Metal _____
Other _____

Weld Metal Thickness _____

ELECTRICAL CHARACTERISTICS (QW-409)
Current _____
Polarity _____
Amps _____ Volts _____
Tungsten Electrode Size _____
Other _____

POSITIONS (QW-405)
Position of Groove _____
Welding Progression (Uphill, Downhill) _____
Other _____

PREHEAT (QW-406)
Preheat Temp. _____
Interpass Temp. _____
Other _____

TECHNIQUE (QW-410)
Travel Speed _____
String or Weave Bead _____
Oscillation _____
Multiple or Single Pass (per side) _____
Single or Multiple Electrodes _____
Other _____

Figure 14.3a   PQR format

QW 483 PQR (Back)

PQR No. __SMAW-1

TENSILE TEST (QW-150)

Specimen No	Width	Thickness	Area	Ultimate Total load lb	Ultimate Unit Stress psi	Type of Failure & Location

GUIDED- BEND TESTS (QW-160)

Type and Figure No	Result

TOUGHNESS TESTS (QW-170)

Specimen No	Notch Location	Specimen Size	Test Temp	Impact Values			Drop Weight Break (Y/N)
				Ft-lb	% Shear	Mils	

Comments _____

FILLET WELD TEST (QW-180)

Result – Satisfactory? : Yes____ No_____   Penetration into Parent Metal? : Yes____ No_____

Macro – Results _____

OTHER TESTS

Type of Test _____

Deposit Analysis _____

Other _____

Welder's Name _____ Clock No. _____ Stamp No. _____

Tests conducted by: _____ Laboratory Test No._____

We certify that the statements in this record are correct and that the tests welds were prepared, welded, and tested in accordance with the requirements of Section IX of the ASME Code.

Manufacturer _____

Date _____   By _____

(Detail of record of tests are illustrative only and may be modified to conform to the type and number of tests required by the Code.)

Figure 14.3b   PQR format

Welding Qualifications and ASME IX

14.2.1 WPSs and PQRs: ASME IX section QW-250

We will now look at the ASME IX code rules covering WPSs and PQRs. The code section splits the variables into three groups:

- Essential variables
- Non-essential variables
- Supplementary variables

These are listed on the WPS for each welding process. ASME IX section QW-250 lists the variables that must be specified on the WPS and PQR for each process. Note how this is a very long section of the code, consisting mainly of tables covering the different welding processes. There are subtle differences between the approaches to each process, but the guiding principles as to what is an essential, non-essential and supplementary variable are much the same.

14.2.2 ASME IX welding documentation formats

The main welding documents specified in ASME IX have examples in non-mandatory appendix B section QW-482. Strangely, these are not included in the API 510 exam code document package but fortunately two of them, the WPS and PQR, are repeated in API 577 (have a look at them in API 577 appendix C). Remember that the actual format of the procedure sheets is not mandatory, as long as the necessary information is included.

The other two that are in ASME IX non-mandatory appendix B (the WPQ and Standard Weld Procedure Specification (SWPS)) are not given in API 577 and are therefore a bit peripheral to the API 510 exam syllabus.

14.3 Welding documentation reviews: the exam questions

The main thrust of the API 510 ASME IX questions is based on the requirement to review a WPS and its qualifying PQR, so these are the documents that you must become familiar with. The review will be subject to the following limitations (to make it simpler for you):

- The WPS and its supporting PQR will contain only *one* welding process.
- The welding process will be SMAW, GTAW, GMAW or SAW and will have only one filler metal.
- The base material P group number will be either P1, P3, P4, P5 or P8.

Base materials are assigned P-numbers in ASME IX to reduce the amount of procedure qualifications required. The P-number is based on material characteristics like weldability and mechanical properties. S-numbers are the same idea as P-numbers but deal with piping materials from ASME B31.3.

14.3.1 WPS/PQR review questions in the exam

The API 510 certification exam requires candidates to review a WPS and its supporting PQR. The format of these will be based on the sample documents contained in annex B of ASME IX. Remember that this annex B is not contained in your code document package; instead, you have to look at the formats in API 577 appendix B, where they *are* shown (they are exactly the same).

The WPS/PQR documents are designed to cover the parameters/variables requirements of the SMAW, GTAW, GMAW and SAW welding processes. The open-book questions on these documents in the API exam, however, only contain *one* of those welding processes. This means that there will be areas on the WPS and PQR documents that will be left unaddressed, depending on what process is used. For example, if GTAW welding is *not* specified then the details of tungsten electrode size and type will not be required on the WPS/PQR.

In the exam questions, you will need to understand the variables to enable you to determine if they have been correctly addressed in the WPS and PQR for any given process.

Welding Qualifications and ASME IX

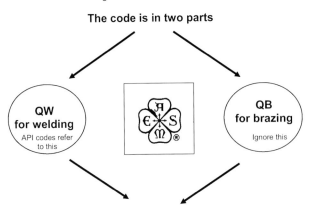

Figure 14.4 The ASME IX numbering system

14.3.2 Code cross-references

One area of ASME IX that some people find confusing is the numbering and cross-referencing of paragraphs that takes place throughout the code. Figure 14.4 explains how the ASME IX numbering system works.

14.4 ASME IX article I

Article 1 contains less technical 'meat' than some of the following articles (particularly articles II and IV). It is more a collection of general statements than a schedule of firm technical requirements. What it does do, however, is cross-reference a lot of other clauses (particularly in article IV), which is where the more detailed technical requirements are contained.

From the API exam viewpoint, most of the questions that can be asked about article I are:

- More suitable to closed-book questions than open-book ones
- Fairly general and 'commonsense' in nature

Don't ignore the content of article I. Read the following summaries through carefully but treat article I more as a lead-in to the other articles, rather than an end in itself.

Section QW-100.1
This section tells you five things, all of which you have met before. There should be nothing new to you here. They are:

- A Weld Procedure Specification (WPS) has to be qualified (by a PQR) by the manufacturer or contractor to determine that a weldment meets its required mechanical properties.
- The WPS specifies the conditions under which welding is performed and these are called welding 'variables'.
- The WPS must address the essential and non-essential variables for each welding process used in production.
- The WPS must address the supplementary essential variables if notch toughness testing is required by other code sections.
- A Procedure Qualification Record (PQR) will document the welding history of the WPS test coupon and record the results of any testing required.

Section QW-100.2
A welder qualification (i.e. the WPQ) is to determine a welder's ability to deposit sound weld metal or a welding operator's mechanical ability to operate machine welding equipment.

Welding Qualifications and ASME IX

14.5 Section QW-140 types and purposes of tests and examinations

Section QW-141: mechanical tests

Mechanical tests used in procedure or performance qualification are as follows:

- **QW-141.1: tension tests** (see Fig. 14.5). Tension tests are used to determine the strength of groove weld joints.
- **QW-141.2: guided-bend tests** (see Fig. 14.6). Guided-bend tests are used to determine the degree of soundness and ductility of groove-weld joints.
- **QW-141.3: fillet-weld tests**. Fillet weld tests are used to

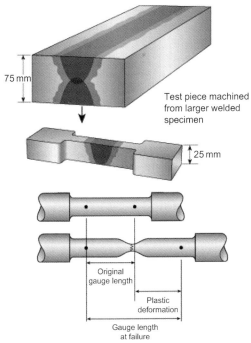

Figure 14.5 Tension tests

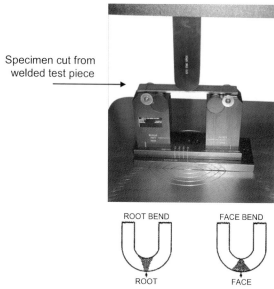

Figure 14.6 Guided bend tests

determine the size, contour and degree of soundness of fillet welds.

- **QW-141.4: notch-toughness tests**. Tests are used to determine the notch toughness of the weldment.

14.6 ASME IX article II

Article II contains hard information about the content of the WPS and PQRs and how they fit together. In common with article I, it cross-references other clauses (particularly in article IV). From the API examination viewpoint there is much more information in here that can form the basis of open-book questions, i.e. about the reviewing of WPS and PQR. ASME IX article II is therefore at the core of the API examination requirements.

Welding Qualifications and ASME IX

Section QW-200: general
This gives lists of (fairly straightforward) requirements for the WPS and PQR:

- **QW-200.1** covers the WPS. It makes fairly general 'principle' points that you need to understand (but not remember word-for-word).
- **QW-200.2** covers the PQR again. It makes fairly general 'principle' points that you need to understand (but not remember word-for-word).
- **QW-200.3: P-numbers**. P-numbers are assigned to base metals to reduce the number of welding procedure qualifications required. For steel and steel alloys, *group* numbers are assigned additionally to P-numbers for the purpose of procedure qualification where notch-toughness requirements are specified.

Now try these familiarization questions, using ASME IX articles I and II to find the answers.

14.7 ASME IX articles I and II familiarization questions

Q1. ASME IX section QW-153: acceptance criteria – tension tests
Which of the following is a true statement on the acceptance criteria of tensile tests?

(a) They must never fail below the UTS of the base material ☐
(b) They must fail in the base material. ☐
(c) They must not fail more than 5 % below the minimum UTS of the base material ☐
(d) They must fail in the weld metal otherwise they are discounted ☐

Q2. ASME IX section QW-200 PQR
A PQR is defined as?

(a) A record supporting a WPS ☐
(b) A record of the welding data used to weld a test coupon ☐

(c) A Procedure Qualification Record ☐
(d) A Provisional Qualification Record ☐

Q3.　ASME IX section QW-200.2 (b)
Who certifies the accuracy of a PQR?

(a) The Authorized Inspector before it can be used ☐
(b) The Manufacturer or his designated subcontractor ☐
(c) An independent third-party organization ☐
(d) Only the Manufacturer or Contractor ☐

Q4.　ASME IX section QW-200.3
What is a P-number?

(a) A number assigned to base metals ☐
(b) A procedure unique number ☐
(c) A number used to group similar filler material types ☐
(d) A unique number designed to group ferrous materials ☐

Q5.　ASME IX section QW-200.3
What does the assignment of a group number to a P-number indicate?

(a) The base material is non-ferrous ☐
(b) Post-weld heat treatment will be required ☐
(c) The base material is a steel or steel alloy ☐
(d) Notch toughness requirements are mandatory ☐

Q6.　ASME IX section QW-202.2 types of test required
What types of mechanical tests are required to qualify a WPS on full penetration groove welds with no notch toughness requirement?

(a) Tension tests and guided bend tests ☐
(b) Tensile tests and impact tests ☐
(c) Tensile, impact and nick break tests ☐
(d) Tension, side bend and macro tests ☐

Q7.　ASME IX section QW-251.1
The 'brief of variables' listed in tables QW-252 to QW-265 reference the variables required for each welding process. Where can the complete list of variables be found?

(a) In ASME B31.3 ☐
(b) In ASME IX article IV ☐
(c) In API 510 ☐
(d) In ASME IX article V ☐

Welding Qualifications and ASME IX

Q8. ASME IX section QW-251.2
What is the purpose of giving base materials a P-number?

(a) It makes identification easier ☐
(b) It reduces the number of welding procedure qualifications required ☐
(c) It shows they are in pipe form ☐
(d) It indicates the number of positions it can be welded in ☐

Q9. ASME IX section QW-251.2
A welder performance test is qualified using base material with an S-number. Which of the following statements is true?

(a) Qualification using an S-number qualifies corresponding S-number materials only ☐
(b) Qualification using an S-number qualifies corresponding F-number materials ☐
(c) Qualification using an S-number qualifies corresponding P-number materials only ☐
(d) Qualification using an S-number qualifies *both* P-number and S-number materials ☐

Q10. ASME IX section QW-253
Which of the following would definitely not be a variable consideration for the SMAW process?

(a) Filler materials ☐
(b) Electrical characteristics ☐
(c) Gas ☐
(d) PWHT ☐

14.8 ASME IX article III

Remember that WPQs are specific to the *welder*. While the content of this article is in the API 510 syllabus it is fair to say that it commands less importance than articles II (WPSs and PQRs and their relevant QW-482 and QW-483 format forms) and article IV (welding data).

Section QW-300.1
This article lists the welding processes separately, with the essential variables that apply to welder and welding operator performance qualifications. The welder qualification is limited by the essential variables listed in QW-350, and

defined in article IV *Welding data*, for each welding process. A welder or welding operator may be qualified by radiography of a test coupon or his initial production welding, or by bend tests taken from a test coupon.

Look at these tables below and mark them with post-it notes:

- Table QW-353 gives SMAW essential variables for welder qualification.
- Table QW-354 gives SAW essential variables for welder qualification.
- Table QW-355 gives GMAW essential variables for welder qualification.
- Table QW-356 gives GTAW essential variables for welder qualification.

Section QW-351: variable for welders (general)
A welder needs to be requalified whenever a change is made in one or more of the essential variables listed for each welding process. The limits of deposited weld metal thickness for which a welder will be qualified are dependent upon the thickness of the weld deposited with each welding process, exclusive of any weld reinforcement.

In production welds, welders may not deposit a thickness greater than that for which they are qualified.

14.9 ASME IX article IV

Article IV contains core data about the welding variables themselves. Whereas article II summarizes which variables are essential/non-essential/supplementary for the main welding processes, the content of article IV explains what the variables actually *are*. Note how variables are subdivided into *procedure* and *performance* aspects.

Section QW-401: general
Each welding variable described in this article is applicable as an essential, supplemental essential or non-essential variable for procedure qualification when referenced in QW-250 for

Welding Qualifications and ASME IX

each specific welding process. Note that a change from one welding process to another welding process is an essential variable and requires requalification.

Section QW-401.1: essential variable (procedure)
This is defined as a change in a welding condition that will affect the mechanical properties (other than notch toughness) of the weldment (for example, change in P-number, welding process, filler metal, electrode, preheat or post-weld heat treatment, etc.).

Section QW-401.2: essential variable (performance)
A change in a welding condition that will affect the ability of a welder to deposit sound weld metal (such as a change in welding process, electrode F-number, deletion of backing, technique, etc.).

Section QW-401.3: supplemental essential variable (procedure)
A change in a welding condition that will affect the notch-toughness properties of a weldment (e.g. change in welding process, uphill or downhill vertical welding, heat input, preheat or PWHT, etc.).

Section QW-401.4: non-essential variable (procedure)
A change in a welding condition that will not affect the mechanical properties of a weldment (such as joint design, method of back-gouging or cleaning, etc.).

Section QW-401.5
The welding data include the welding variables grouped as follows:

- QW-402 joints
- QW-403 base metals
- QW-404 filler metal
- QW-405 position
- QW-406 preheat
- QW-407 post-weld heat treatment

- QW-408 gas
- QW-409 electrical characteristics
- QW-410 technique

Section QW-420.1: P-numbers
P-numbers are groupings of base materials of similar properties and usability. This grouping of materials allows a reduction in the number of PQRs required. Ferrous P-number metals are assigned a group number if notch toughness is a consideration.

Section QW-420.2: S-numbers (non-mandatory)
S-numbers are similar to P-numbers but are used on materials not included within ASME BPV code material specifications (section II). There is no mandatory requirement that S-numbers have to be used, but they often are. Note these two key points:

- For WPS a P-number qualifies the same S-number but not vice versa.
- For WPQ a P-number qualifies the same S-number and vice versa.

Section QW-430: F-numbers
The F-number grouping of electrodes and welding rods is based essentially on their usability characteristics. This grouping is made to reduce the number of welding procedure and performance qualifications, where this can logically be done.

Section QW-432.1
Steel and steel alloys utilize F-1 to F-6 and are the most commonly used ones.

Section QW-492: definitions
QW-492 contains a list of definitions of the common terms relating to welding and brazing that are used in ASME IX.

Try these ASME IX articles III and IV familiarization

Welding Qualifications and ASME IX

questions. You will need to refer to your code to find the answers.

14.10 ASME IX articles III and IV familiarization questions

Q1. ASME IX section QW-300
What does ASME IX article III contain?

(a) Welding Performance Qualification requirements ☐
(b) A list of welding processes with essential variables applying to WPQ ☐
(c) Welder qualification renewals ☐
(d) All of the above ☐

Q2. ASME IX section QW-300.1
What methods can be used to qualify a welder?

(a) By visual and bend tests taken from a test coupon ☐
(b) By visual and radiography of a test coupon or his initial production weld ☐
(c) By visual, macro and fracture test ☐
(d) Any of the above can be used depending on joint type ☐

Q3. ASME IX section QW-301.3
What must a manufacturer or contractor *not* assign to a qualified welder to enable his work to be identified?

(a) An identifying number ☐
(b) An identifying letter ☐
(c) An identifying symbol ☐
(d) Any of the above can be assigned ☐

Q4. ASME IX section QW-302.2
If a welder is qualified by radiography, what is the minimum length of coupon required?

(a) 12 inches (300 mm) ☐
(b) 6 inches (150 mm) ☐
(c) 3 inches (75 mm) ☐
(d) 10 inches (250 mm) ☐

Q5. ASME IX section QW-302.4
What areas of a pipe test coupon require visual inspection for a WPQ?

(a) Inside and outside of the entire circumference ☐

(b) Only outside the surface if radiography is to be used ☐
(c) Only the weld metal on the face and root ☐
(d) Visual inspection is not required for pipe coupons ☐

Q6. ASME IX section QW-304

For a WPQ, which of the following welding processes can *not* have groove welds qualified by radiography?

(a) GMAW (short-circuiting transfer mode) ☐
(b) GTAW ☐
(c) GMAW (globular transfer mode) ☐
(d) They can all be qualified by radiography ☐

Q7. ASME IX section QW-322

How long does a welder's performance qualification last if he has not been involved in production welds using the qualified welding process?

(a) 6 months ☐
(b) 2 years ☐
(c) 3 months ☐
(d) 6 weeks ☐

Q8. ASME IX section QW-402 joints

A welder qualified in a single welded groove weld with backing must requalify if:

(a) He must now weld without backing ☐
(b) The backing material has a nominal change in its composition ☐
(c) There is an increase in the fit-up gap beyond that originally qualified ☐
(d) Any of the above occur ☐

Q9. ASME IX section QW-409.8

What process requires the electrode wire feed speed range to be specified?

(a) SMAW ☐
(b) SAW ☐
(c) GMAW ☐
(d) This term is not used in ASME IX ☐

Q10. ASME IX section QW-416
Which of the following variables would not be included in a WPQ?

(a) Preheat ☐
(b) PWHT ☐
(c) Technique ☐
(d) All of them ☐

14.11 The ASME IX review methodology

One of the major parts of all the API in-service inspection examinations is the topic of weld procedure documentation review. In addition to various 'closed-book' questions about welding processes and techniques, the exams always include a group of 'open-book' questions centred around the activity of checking a Weld Procedure Specification (WPS) and Procedure Qualification Record (PQR).

Note the two governing principles of API examination questions on this subject:

- The PQR and WPS used in exam examples will only contain one welding process and filler material.
- You need only consider essential and non-essential variables (you can ignore supplementary variables).

The basic review methodology is divided into five steps (see Fig. 14.7). Note the following points to remember as you go through the checklist steps of Fig. 14.7:

- The welding *process* is an *essential* variable and is likely to be SMAW, GTAW, GMAW or SAW.
- Non-essential variables do not have to be recorded in the PQR (but may be at the manufacturer's discretion) and must be addressed in the WPS.
- Information on the PQR will be *actual values* used whereas the WPS may contain a *range* (e.g. the base metal *actual* thickness shown in a PQR may be $\frac{1}{2}$ in, while the base metal thickness *range* in the WPS may be $\frac{3}{16}$ in–1 in).
- The process variables listed in tables QW-252 to QW-265

STEP 1: variables table

- Find the relevant 'brief of variables' table in article 2 of ASME IX for the specified welding process (for example QW-253 for SMAW). This table shows the relevant essential and non-essential variables for the welding process.

STEP 2: PQR check

- Check that the 'editorial' information at the beginning and at the end of the PQR form is filled in.
- Check that all the relevant **essential** variables are addressed on the PQR and highlight any that are not.

STEP 3: WPS check

- Check that the editorial information at the beginning of the WPS form is filled in and agrees with the information on the PQR.
- Check that all the relevant **essential** variables are addressed on the WPS and highlight any that are not.
- Check that all the relevant **non-essential** variables are addressed on the WPS and highlight any that are not.

STEP 4: range of qualification

- Check that the **range of qualification** for each **essential variable** addressed in the PQR is correct and has been correctly stated on the WPS.

STEP 5: number of tensile and bend tests

- Check that the correct type and number of tensile and bend tests have been carried out and recorded on the PQR.
- Check that the tensile/bend test results are correct.

Figure 14.7 The ASME IX WPS/PQR review methodology

are referred to as the 'brief of variables' and must *not* be used on their own. You *must* refer to the full variable requirements referenced in ASME IX article 4 otherwise you will soon find yourself in trouble.
- The base material will be either P-1, P-3, P-4, P-5 or P-8 (base materials are assigned P-numbers in ASME IX to reduce the amount of procedure qualifications required).

14.12 ASME IX WPS/PQR review: worked example

The following WPS/PQR is for an SMAW process and contains typical information that would be included in an exam question. Work through the example and then try the questions at the end to see if you have understood the method.

Figures 14.8 and 14.9 show the WPS and PQR for an SMAW process. Typical questions are given, followed by their answer and explanation.

Quick Guide to API 510

WPS

Company Name: MET Ltd By: S. Hughes
Welding Procedure Specification No: SMAW-1 Date: 01/04/2006 Supporting PQR No: SMAW-1
Revision No: 0 Date: 01/04/06

Welding Process(es): SMAW Type(s): Manual
Automatic, Manual, machine or Semi Automatic

JOINTS (QW-402)

Details

Joint Design: Single Vee Butt
Backing (Yes) _____ (No) X
Backing Material (Type) _____

See production drawing

Refer to both backing and retainers

☐ Metal ☐ Nonfusing Metal

☐ Nonmetallic ☐ Other

Sketches, Production Drawings, Weld Symbols or Written Description should show the general arrangement of the parts to be welded. Where applicable, the root spacing and the details of weld groove may be specified. At the option of the manufacturer, sketches may be attached to illustrate joint design, weld layers and bead sequence, eg for notch toughness procedures, for multiple process procedures etc

BASE METALS (QW-403)

P-No. 1 Group No. 2 to P-No. 1 Group No. 2
OR
Specification type and grade _____
to
Specification type and grade _____
OR
Chemical Analysis and Mechanical properties _____
to
Chemical Analysis and Mechanical properties _____

Thickness Range:
Base Metal: Groove 1/16" – 2" Fillet _____
Pipe Diameter range: Groove All Fillet All
Other _____

FILLER METALS (QW-404) Each base metal-filler metal combination should be recorded individually

Spec. No (SFA): SFA 5.1
AWS No (Class): E7016
F No: 6
A-No: 4
Size of filler metals: All
Weld Metal
Thickness range
  Groove: All
  Fillet: All
Electrode-Flux (Class): N/A
Flux Trade Name: N/A
Consumable Insert: N/A
Other

Figure 14.8a SMAW worked example (WPS)

Welding Qualifications and ASME IX

WPS (Back)

WPS No _SMAW-1___ Rev ___0___

POSITIONS (QW-405)
Position(s) of Groove ___All___
Welding Progression: Up __Yes__ Down __Yes__
Position(s) of Fillet ___All___

POSTWELD HEAT TREATMENT (QW407)
Temperature Range ___None___
Time Range ___None___

PREHEAT (QW-406)
Rate
Preheat Temp. Min ___None___
Interpass Temp. Max ___None___
Preheat Maintenance ___None___

(Continuous or special heating, where applicable, should be recorded)

GAS (QW-408)
Percent composition
Gas(es)   (Mixture)   Flow
Shielding ___None___
Trailing ___None___
Backing ___None___

ELECTRICAL CHARACTERISTICS (QW-409)
Current AC or DC ___DC___   Polarity ___Reverse___
Amps (Range) ___110-120___   Volts (Range) ___12-20___
(Amps and volts range should be recorded for each electrode size, position, and thickness, etc. this information may be listed in a tabular form similar to that shown below).

Tungsten Electrode Size and Type ___N/A___
(Pure Tungsten, 2% Thoriated, etc)
Mode of Metal Transfer for GMAW ___N/A___
(Spray arc, short circuiting arc, etc)
Electrode Wire feed speed range ___N/A___

TECHNIQUE (QW-410)
String or Weave Bead ___Both___
Orifice or Gas Cup Size ___N/A___
Initial and Interpass Cleaning (Brushing, Grinding, etc) ___Brushing, grinding___
Method of Back Gouging ___None___
Oscillation ___N/A___
Contact Tube to Work Distance ___N/A___
Multiple or Single Pass (per side) ___Multiple pass – no pass greater than ¼"___
Multiple or Single Electrodes ___Multiple___
Travel Speed (Range) ___10 IPM___
Peening ___Allowed___
Other

Weld Layer(s)	Process	Filler Metal		Current			Travel Speed Range	Other (remarks, comments, hot wire addition, technique, torch angle etc)
		Class	Diameter	Type Polarity	Amp Range	Volt range		

Figure 14.8b   SMAW worked example (WPS)

PQR

Company Name: MET Ltd
Procedure Qualification Record No.: SMAW-1   Date: 19/03/06
WPS No: SMAW-1
Welding Process(es): SMAW
Types (Manual, Automatic, Semi-Auto): Manual

JOINTS (QW-402)

Single Vee Groove, 60 degree included angle. No backing

Groove Design of Test Coupon
(For combination qualifications, the deposited weld metal thickness that shall be recorded for each filler metal or process used)

BASE METALS (QW-403)
Material spec.: SA672
Type or Grade: B70
P-No: 1 to P-No: 1
Thickness of test coupon: 1"
Diameter of test coupon: 36"
Other:

POSTWELD HEAT TREATMENT (QW-407)
Temperature: None
Time: None
Other:

GAS (QW-408)
Percent composition
Gas(es) (Mixture) Flow Rate
Trailing: None
Backing: None
Backing: None

FILLER METALS (QW-404)
SFA Specification: 5.1
AWS Classification: E7018
Filler metal F-No: 4
Weld Metal Analysis A-No: 1
Size of Filler Metal: 1/8"
Other:
Weld Metal Thickness: 1"

ELECTRICAL CHARACTERISTICS (QW-409)
Current: Direct
Polarity: Reverse
Amps: 100   Volts: 10
Tungsten Electrode Size: N/A
Other:

POSITIONS (QW-405)
Position of Groove: 3G
Welding Progression (Uphill, Downhill):
Other:

PREHEAT (QW-406)
Preheat Temp.: 200°F
Interpass Temp.: 650°F
Other:

TECHNIQUE (QW-410)
Travel Speed: 25 IPM
String or Weave Bead: String
Oscillation: N/A
Multiple or Single Pass (per side): multiple
Single or Multiple Electrodes: Single
Other:

Figure 14.9a   SMAW worked example (PQR)

Welding Qualifications and ASME IX

PQR (Back)
PQR No. __SMAW-1__

TENSILE TEST (QW-150)

Specimen No	Width	Thickness	Area	Ultimate Total load lb	Ultimate Unit Stress psi	Type of Failure & Location
T-1	0.750	0.985	0.7387	54100	73236	BF/WM
T-2	0.751	0.975	0.6253	40000	63969	BF/WM

GUIDED- BEND TESTS (QW-160)

Type and Figure No	Result
QW-462.2- FACE	Opening 1/16" long – Acceptable
QW-462.2- ROOT	Acceptable

TOUGHNESS TESTS (QW-170)

Specimen No	Notch Location	Specimen Size	Test Temp	Ft-lb	% Shear	Mils	Drop Weight Break (Y/N)

Comments _____

FILLET WELD TEST (QW-180)

Result – Satisfactory? : Yes ____ No ____   Penetration into Parent Metal? : Yes ____ No ____

Macro – Results _____

OTHER TESTS

Type of Test _____

Deposit Analysis _____

Other _____

Welder's Name __Richard Easton__   Clock No. _____   Stamp No. __RE2__

Tests conducted by: __LAB Ltd__   Laboratory Test No. __LAB01__

We certify that the statements in this record are correct and that the tests welds were prepared, welded, and tested in accordance with the requirements of Section IX of the ASME Code.

Manufacturer __MET Ltd__

Date __19/03/06__   By __S. Hughes__

Figure 14.9b   SMAW worked example (PQR)

Step 1: variable table

Q1 (WPS). The base metal thickness range shown on the WPS:
(a) Is correct ☐
(b) Is wrong – it should be $\frac{1}{16}$ in–$1\frac{1}{2}$ in ☐
(c) Is wrong – it should be $\frac{3}{16}$ in–2 in (QW 451.1) ☐
(d) Is wrong – it should be $\frac{3}{8}$ in–1 in ☐

The welding process is SMAW; therefore the brief of variables used will be those in table QW-253. Look at table QW-253 and check the brief of variables for base metals (QW-403). Note that QW-403.8 specifies that 'change' of thickness T qualified as an essential variable and therefore the base material thickness must be addressed on the PQR. When we read QW-403.8 in section IV we see that it refers us to QW-451 for the thickness range qualified. Thus:

- The PQR tells us under base metals (QW-403) the coupon thickness $T = 1$ inch.
- QW-451.1 tells us that for a test coupon of thickness $\frac{3}{4}$–$1\frac{1}{2}$ inch the base material range qualified on the WPS is $\frac{3}{16}$ inch to $2T$ (therefore $2T = 2$ inches).

The correct answer must therefore be (c).

Q2 (WPS). The deposited weld metal thickness:
(a) Is correct ☐
(b) Is wrong – it should be 'unlimited' ☐
(c) Is wrong – it should be 8 in maximum ☐
(d) Is wrong – it should be 2 in maximum (QW-451.1) ☐

Look at table QW-253 and note how QW-404.30 'change in deposited weld metal thickness t' is an essential variable (and refers to QW-451 for the maximum thickness qualified); therefore weld metal thickness must be addressed in the PQR. Thus:

- PQR under QW-404 filler states weld metal thickness $t = 1$ inch.

Welding Qualifications and ASME IX

- QW-451 states if $t \geq \frac{3}{4}$ in then maximum qualified weld metal thickness $= 2T$ where $T =$ base metal thickness.

The correct answer must therefore be (d).

Q3 (WPS): check of consumable type. The electrode change from E7018 on the PQR to E7016 on the WPS:
(a) Is acceptable (QW-432) ☐
(b) Is unacceptable – it can only be an E7018 on the WPS ☐
(c) Is acceptable – provided the electrode is an E7016 A1 ☐
(d) Is unacceptable – the only alternate electrode is an E6010 ☐

Note how QW-404.4 shows that a change in F-number from table QW-432 is an essential variable. This is addressed on the PQR, which shows the E7018 electrode as an F- 4. Table QW-432 and the WPS both show the E7016 electrode is also an F-4.

The correct answer must therefore be (a).

Q4 (WPS): preheat check. The preheat should read:
(a) 60 °F minimum ☐
(b) 100 °F minimum ☐
(c) 250 °F minimum ☐
(d) 300 °F minimum ☐

QW-406.1 shows that a decrease of preheat $>$ 100 °F (55 °C) is defined as an essential variable.

The PQR shows a preheat of 200 °F, which means the minimum shown on the WPS must be 100 °F and not 'none' as shown.

The correct answer must therefore be (b).

Q5 (PQR): Check tensile test results. The tension tests results are:
(a) Acceptable ☐
(b) Unacceptable – not enough specimens ☐
(c) Unacceptable – UTS does not meet ASME IX (QW-422) or 70 ksi ☐
(d) Unacceptable – specimen width is incorrect ☐

Note how the tensile test part of the PQR directs you to QW-150. On reading this section you will notice that it directs you to the tension test acceptance criteria in QW-153. This says that the minimum procedure qualification tensile values are found in table QW/QB-422. Checking through the figures for material SA-672 grade B70 shows a minimum specified tensile value of 70 ksi (70 000 psi) but the PQR specimen T-2 shows a UTS value of 63 969 psi.

The correct answer must therefore be (c).

Q6 (PQR). The bend test results are:
(a) Acceptable ☐
(b) Unacceptable – defect size greater than permitted ☐
(c) Unacceptable – wrong type and number of specimens (QW-450) ☐
(d) Unacceptable – incorrect figure number – should be QW-463.2 ☐

The PQR directs you to QW-160 for bend tests. For API exam purposes the bend tests will be *transverse* tests. Note these important sections covering bend tests:

- QW-163 gives acceptance criteria for bend tests.
- QW-451 contains PQR thickness limits and test specimen requirements.
- QW-463.2 refers to performance qualifications.

Check of acceptance criteria
From QW-163, the $\frac{1}{16}$ inch defect is acceptable so answer (b) is incorrect. QW-463.2 refers to *performance qualifications* so answer (d) is incorrect.

Check of test specimen requirements
QW-451 contains the PQR thickness limits and test specimen requirements. Consulting this, we can see that for this material thickness (1 in) there is a requirement for *four side bend* tests.

Welding Qualifications and ASME IX

*Therefore the correct answer is **(c)**.*

Q7 (PQR): validation of PQRs. To be 'code legal' the PQR must be:
(a) **Certified (QW-201)** ☐
(b) Notarized ☐
(c) Authorized ☐
(d) Witnessed ☐

The requirements for certifying of PQRs is clearly shown in QW-201. Note how it says *'the manufacturer or contractor shall certify that he has qualified...'*

*The correct answer must therefore be **(a)**.*

Q8 (WPS/PQR): check if variables shown on WPS/PQR. Essential variable QW-403.9 has been:
(a) Correctly addressed on the WPS ☐
(b) Incorrectly addressed on the WPS ☐
(c) Not addressed on the PQR ☐
(d) **Both (b) and (c)** ☐

Note how QW-253 defines QW-403.9 '*t*-pass' as an essential variable. It must therefore be included on the PQR *and* WPS. Note how in the example it has been addressed on the WPS (under the QW-410 technique) but has not been addressed on the PQR.

*The correct answer must therefore be **(d)**.*

Q9 (PQR): variables shown on WPS/PQR. The position of the groove weld is:
(a) Acceptable as shown ☐
(b) Unacceptable – it is an essential variable not addressed ☐
(c) **Unacceptable – position shown is not for pipe (QW-461.4)** ☐
(d) Both (b) and (c) ☐

Remember that weld positions are shown in QW-461. They are not an essential variable however, so the weld position is not required to be addressed on the PQR. If it is (optionally) shown on the PQR it needs to be checked to make sure it is correct. In this case the position shown refers to the test position of the plate, rather than the pipe.

The correct answer must therefore be (c).

Q10 (PQR/WPS): variables. The PQR shows 'string beads but WPS shows 'both' string and weave beads. This is:
(a) Unacceptable – does not meet code requirements ☐
(b) Acceptable – meets code requirements (non-essential variable QW-200.1c) ☐
(c) Acceptable – if string beads are only used on the root ☐
(d) Acceptable – if weave beads are only used on the cap ☐

For SMAW, the type of weld bead used is not specified under QW-410 as an essential variable. This means it is a *non-essential* variable and is not required in the PQR (but remember it can be included by choice). QW-200.1 (c) permits changes to non-essential variables of a WPS as long as it is recorded. It is therefore acceptable to specify a string bead in the PQR but record it as 'string and weave' in the WPS.

The correct answer must therefore be (b).

Chapter 15

The NDE Requirements of API 510 and API 577

15.1 NDE Requirements of API 510 and API 577

In many ways, the API 510 *Body of Knowledge* contains a patchwork quilt of NDE requirements. Figure 15.1 shows the situation; ASME V, VIII and IX all contain NDE requirements related to their own new construction focus while API 510, 577 and 572 supplement this with their own requirements related to in-service inspection, and then repair. API 510, remember, retains its position as the 'override' code – taking priority over the others wherever conflict exists (and there are a few such areas).

Other sections of this book cover the requirements of ASME V, VIII and IX in some detail. These are by far the longest sections, as you would expect, as they come from a fully blown construction code. API 577, being a Recommended Practice (RP) document rather than a formal code, takes an almost 'textbook' approach. It contains an extremely diverse, and in places quite deep, coverage of metallurgy, welding, NDE and almost everything else – in many areas far too detailed to be included in the API exam.

15.2 API 510 NDE requirements

Surprisingly, API 510 itself does not contain much direct information on NDE at all. What little it does contain is fragmented throughout various chapters of the code in snippets, rather than in a separate chapter. This has three main results:

- It is more difficult to find, as it is not contained in one place.
- The 'snippet' form makes it more suitable for closed-book exam questions.

Quick Guide to API 510

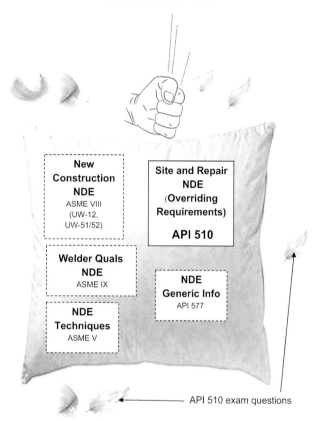

Figure 15.1 The API 510 patchwork of NDE requirements

- These questions tend to owe more to examination convenience rather than the value of inspection knowledge they contain.

In fairness, API 510 does not pretend to be an NDE-orientated code. It is happy to concentrate more on what to do with the results of NDE activities, leaving the description

The NDE Requirements of API 510 and API 577

of the technique themselves to other related codes such as ASME V.

15.2.1 Links to API 571 and API 577

API 571, covering damage mechanisms, contains a lot of information on the NDE technique. It relates these to their suitability for finding the *results* of various damage mechanisms. This document contains a lot of technical opinion, which means that it has to be judgemental on which NDE techniques can and can't find specific damage mechanisms and defects. You may find, therefore, that you do not actually *agree* with all of it.

Referring back to API 510, Fig. 15.2 shows some specific sections that contain NDE requirements. Some of these are fact and some are API opinion, but all can contain valid API 510 exam questions. Note how they are all fairly thin on detail, consisting mainly of short statements rather than elaborate technical argument or justifications.

Quick Guide to API 510

API 510 SECTION	SUBJECT	API 510's VIEW
Definition 3.1	What is a *defect*?	A defect is an indication that exceeds the applicable acceptance criteria.
Definition 3.2	So what is an *indication*?	An indication is just something found by NDE – it may be a defect or it may not.
4.2.4	Responsibilities of the inspector	It is the inspector's job to make sure that NDE meets API 510 requirements.
4.2.5	Who actually does the NDE?	API 510 calls NDE technicians or operatives *examiners*.
5.1.2	Where are NDE activities specified?	In the inspection plan (see 5.1.2(d)).
5.5.3.2	On-stream inspections	Non-intrusive (meaning NDE) examinations can be used in some situations to replace vessel internal examination.
5.7.1	Choice of NDE technique	This section (a) to (j) contains multiple value-judgements on which NDE technique is best for finding what – a common source of exam questions.
5.7.1.2	Shear wave operator qualification	It is the plant owner/user's job to specify that shear wave 'examiners' need adequate qualification.
5.7.2.1	Thickness measurement methods	A, B or C scan UT are suitable for numerous measurement activities.
5.7.2.4	NDE inaccuracies	NDE techniques all have measurement inaccuracies.
8.1.2.1	Approval of repairs	NDE of repairs must be approved by the inspector.

The NDE Requirements of API 510 and API 577

8.1.5.4.4	Repairs to stainless steel overlay	Base metal is to be checked by UT to detect post-weld cracking.
8.1.7	NDE of repair welds	PT/MT should be performed on weld preparations before welding.
8.1.7.3	NDE of repair welds	Repairs require RT (or equivalent) as per the original construction code that was used for the new vessel.
8.1.8	Weld inspection of brittle components	NDE is required to find cracks and notches.
8.2.1	Re-rating	NDE is an acceptable substitute for pressure testing in proving vessel integrity.
Annex B3.2	Inspector recertification	NDE experience may be considered as 'active engagement as an inspector'.

Figure 15.2 Specific NDE requirements of API 510 9th edition

Chapter 16

The NDE Requirements of ASME V

16.1 Introduction

This chapter is to familiarize you with the specific NDE requirements contained in ASME V. ASME VIII references ASME V as the supporting code but only articles 1, 2, 6, 7, 9 and 23 are required for use in the API 510 examination.

These articles of ASME V provide the main detail of the NDE techniques that are referred to in many of the API codes. Note that it is only the *body* of the articles that are included in the API examinations; the additional (mandatory and non-mandatory) appendices that some of the articles have are not examinable. We will now look at each of the articles 1, 2, 6, 7, 9 and 23 in turn.

16.2 ASME V article 1: general requirements

Article 1 does little more than set the general scene for the other articles that follow. It covers the general requirement for documentation procedures, equipment calibration and records, etc., but doesn't go into technique-specific detail. Note how the subsections are annotated with T-numbers (as opposed to I-numbers used for the appendices).

Manufacturer versus repairer
One thing that you may find confusing in these articles is the continued reference to *The Manufacturer*. Remember that ASME V is really a code intended for new manufacture. We are using it in its API 570 context, i.e. when it is used to cover repairs. In this context, you can think of The Manufacturer as *The Repairer*.

Table A-110: imperfections and types of NDE method
This table lists imperfections in materials, components and welds and the suggested NDE methods capable of detecting them. Note how it uses the terminology *imperfection* ...

some of the other codes would refer to these as discontinuities or indications (yes, it is confusing). Note that table A-110 is divided into three types of imperfection:

- Service-induced imperfections
- Welding imperfections
- Product form

We are mostly concerned with the service-induced imperfections and welding imperfections because our NDE techniques are to be used with API 570, which deals with in-service inspections and welding repairs.

The NDE methods in table A-110 are divided into those that are capable of finding imperfections that are:

- Open to the surface only
- Open to the surface or slightly subsurface
- Located anywhere through the thickness examined

Note how article 1 provides very basic background information only. The main requirements appear in the other articles, so API examination questions on the actual content of article 1 are generally fairly rare. If they do appear they will probably be closed book, with a very general theme.

16.3 ASME V article 2: radiographic examination

ASME V article 2 covers some of the specifics of radiographic testing techniques. Note that it does not cover anything to do with the *extent* of RT on pipework, i.e. how many radiographs to take or where to do them (we have seen previously that these are covered in ASME B31.3).

Most of article 2 is actually taken up by details of image quality indicators (IQIs) or penetrameters, and parameters such as radiographic density, geometric unsharpness and similar detailed matters. While this is all fairly specialized, it is fair to say that the subject matter lends itself more to open-book exam questions rather than closed-book 'memory' types of questions.

T-210: scope

This explains that article 2 is used in conjunction with the general requirements of article 1 for the examination of materials including castings and welds.

Note that there are seven mandatory appendices detailing the requirements for other product-specific, technique-specific and application-specific procedures. Apart from appendix V, which is a glossary of terms, do not spend time studying these appendices. Just look at the titles and be aware they exist. The same applies to the three non-mandatory appendices.

T-224: radiograph identification

Radiographs have to contain unique traceable permanent identification, along with the identity of the manufacturer and date of radiograph. The information need not be an image that actually appears on the radiograph itself (i.e. it could be from an indelible marker pen) but usually is.

T-276: IQI (image quality indicator) selection

T-276.1: material

IQIs have to be selected from either the same alloy material group or an alloy material group or grade with less radiation absorption than the material being radiographed.

Remember that the IQI gives an indication of how 'sensitive' a radiograph is. The idea is that the smallest wire visible will equate to the smallest imperfection size that will be visible on the radiograph.

T-276.2: size of IQI to be used (see Fig. 16.1)

Table T-276 specifies IQI selection for various material thickness ranges. It gives the designated hole size (for hole type IQIs) and the essential wire (for wire type IQIs) when the IQI is placed on either the source side or film side of the weld. Note that the situation differs slightly depending on whether the weld has reinforcement (i.e. a weld cap) or not.

The NDE Requirements of ASME V

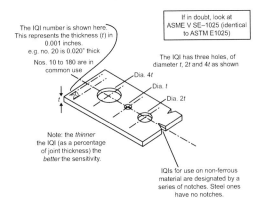

The IQI number is shown here. This represents the thickness (t) in 0.001 inches. e.g. no. 20 is 0.020" thick. Nos. 10 to 180 are in common use

If in doubt, look at ASME V SE–1025 (identical to ASTM E1025)

The IQI has three holes, of diameter t, $2t$ and $4t$ as shown

Note: the *thinner* the IQI (as a percentage of joint thickness) the *better* the sensitivity.

IQIs for use on non-ferrous material are designated by a series of notches. Steel ones have no notches.

Image quality designation is expressed as $(X)-(Y)t$:
(X) is the IQI thickness (t) expressed as a percentage of the joint thickness
$(Y)(t)$ is the hole that must be visible

IQI designation	Sensitivity	Visible hole*
1–2t	1	2t
2–1t	1.4	1t
2–2t	2.0	2t
2–4t	2.8	4t
4–2t	4.0	2t

* The hole that must be visible in order to ensure the sensitivity level shown

T-277.1   ARTICLE 2 – RADIOGRAPHIC EXAMINATION   T-277.2

TABLE T-276
IQI SELECTION

Nominal Single-Wall Material Thickness Range		IQI			
		Source Side		Film Side	
in.	mm	Hole-Type Designation	Wire-Type Essential Wire	Hole-Type Designation	Wire-Type Essential Wire
Up to 0.25, incl.	Up to 6.4, incl.	12	5	10	4
Over 0.25 through 0.375	Over 6.4 through 9.5	15	6	12	5
Over 0.375 through 0.50	Over 9.5 through 12.7	17	7	15	6
Over 0.50 through 0.75	Over 12.7 through 19.0	20	8	17	7
Over 0.75 through 1.00	Over 19.0 through 25.4	25	9	20	8
Over 1.00 through 1.50	Over 25.4 through 38.1	30	10	25	9
Over 1.50 through 2.00	Over 38.1 through 50.8	35	11	30	10
Over 2.00 through 2.50	Over 50.8 through 63.5	40	12	35	11
Over 2.50 through 4.00	Over 63.5 through 101.6	50	13	40	12
Over 4.00 through 6.00	Over 101.6 through 152.4	60	14	50	13
Over 6.00 through 8.00	Over 152.4 through 203.2	80	16	60	14
Over 8.00 through 10.00	Over 203.2 through 254.0	100	17	80	16
Over 10.00 through 12.00	Over 254.0 through 304.8	120	18	100	17
Over 12.00 through 16.00	Over 304.8 through 406.4	160	20	120	18
Over 16.00 through 20.00	Over 406.4 through 508.0	200	21	160	20

Figure 16.1 IQI selection

T-277: use of IQIs to monitor radiographic examination

T-277.1: placement of IQIs

For the best results, IQIs are placed on the *source side* (i.e. nearest the radiographic source) of the part being examined. If inaccessibility prevents hand-placing the IQI on the source side, *it can be placed on the film side* in contact with the part being examined. If this is done, a lead letter 'F' must be placed adjacent to or on the IQI to show it is on the film side. This will show up on the film.

IQI location for welds. Hole type IQIs can be placed adjacent to or on the weld. Wire IQIs are placed on the weld so that the length of the wires is perpendicular to the length of the weld. The identification number(s) and, when used, the lead letter 'F' must not be in the area of interest, except where the geometric configuration of the component makes it impractical.

T-277.2: number of IQIs to be used

At least one IQI image must appear on *each radiograph* (except in some special cases). If the radiographic density requirements are met by using more than one IQI, one must be placed in the lightest area and the other in the darkest area of interest. The idea of this is that the intervening areas are then considered as having acceptable density (a sort of interpolation).

T-280: evaluation of radiographs (Fig. 16.2)

This section gives some quite detailed 'quality' requirements designed to make sure that the radiographs are readable and interpreted correctly.

T-282: radiographic density

These are specific requirements that are based on very well-established requirements used throughout the NDE industry. It gives numerical values of *density* (a specific measured parameter) that have to be met for a film to be considered acceptable.

The NDE Requirements of ASME V

This introduces four parameters that define the 'quality' of a radiograph

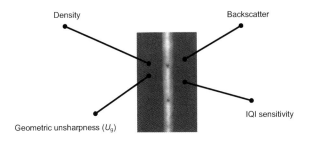

Be prepared to learn about these parameters, and what their values/limits are

Figure 16.2 Evaluation of radiographs

T-282.1: density limitations
This specifies acceptable density limits as follows:

- Single film with X-ray source: density = 1.8 to 4.0
- Single film with gamma-ray source: density = 2.0 to 4.0
- Multiple films: density = 0.3 to 4.0

A tolerance of 0.05 in density is allowed for variations between densitometer readings.

T-283: IQI sensitivity

T-283.1: required sensitivity
In order for a radiograph to be deemed 'sensitive enough' to show the defects of a required size, the following things must be visible when viewing the film:

- For a hole type IQI: the designated hole IQI image and the 2T hole
- For a wire type IQI: the designated wire
- IQI identifying numbers and letters

This is undesirable stray radiation (because it fogs the image)

- A lead letter **B** is attached to the back of the film
- If it appears on the image as a light image on a dark background, the image is unacceptable

Figure 16.3   Backscatter gives an unclear image

T-284: excessive backscatter
Backscatter is a term given to the effect of scattering of the X or gamma rays, leading to an unclear image.

If a light image of the lead symbol 'B' appears on a darker background on the radiograph, protection from backscatter is insufficient and the radiograph is unacceptable. A dark image of 'B' on a lighter background is acceptable (Fig. 16.3).

T-285: geometric unsharpness limitations
Geometric unsharpness is a numerical value related to the 'fuzziness' of a radiographic image, i.e. an indistinct 'penumbra' area around the outside of the image. It is represented by a parameter U_g (unsharpness due to geometry) calculated from the specimen-to-film distance, focal spot size, etc.

Article 2 section T-285 specifies that geometric unsharpness (U_g) of a radiograph shall not exceed the following:

Material thickness, in (mm)	U_g Maximum, in (mm)
Under 2 (50.8)	0.020 (0.51)
2 through 3 (50.8–76.2)	0.030 (0.76)
Over 3 through 4 (76.2–101.6)	0.040 (1.02)
Greater than 4 (101.6)	0.070 (1.78)

In all cases, material thickness is defined as the thickness on which the IQI is chosen.

16.4 ASME V article 6: penetrant testing (PT)

T-620: general
This article of ASME V explains the principle of penetrant testing (PT). We have already covered much of this in API 577, but ASME V article 6 adds some more formal detail.

T-642: surface preparation before doing PT
Surfaces can be in the as-welded, as-rolled, as-cast or as-forged condition and may be prepared by grinding, machining or other methods as necessary to prevent surface irregularities masking indications. The area of interest, and adjacent surfaces within 1 inch (25 mm), need to be prepared and degreased so that indications open to the surface are not obscured.

T-651: the PT techniques themselves
Article 6 recognizes *three penetrant processes*:

- Water washable
- Post-emulsifying (not water based but will wash off with water)
- Solvent removable

The three processes are used in combination with the *two penetrant types* (visible or fluorescent), resulting in a total of six liquid penetrant techniques.

T-652: PT techniques for standard temperatures
For a standard PT technique, the temperature of the penetrant and the surface of the part to be processed must be between 50 °F (10 °C) and 125 °F (52 °C) throughout the examination period. Local heating or cooling is permitted to maintain this temperature range.

T-670: the PT examination technique (see Fig. 16.4)

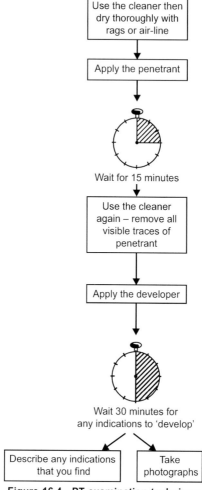

Figure 16.4 PT examination technique

T-671: penetrant application

Penetrant may be applied by any suitable means, such as dipping, brushing or spraying. If the penetrant is applied by spraying using compressed-air type apparatus, filters have to be placed on the upstream side near the air inlet to stop contamination of the penetrant by oil, water, dirt or sediment that may have collected in the lines.

T-672: penetration time

Penetration time is critical. The minimum penetration time must be as required in table T-672 or as qualified by demonstration for specific applications.

Note: While it is always a good idea to follow the manufacturers' instructions regarding use and dwell times for their penetrant materials, table T-672 lays down *minimum* dwell times for the penetrant and developer. These are the minimum values that would form the basis of any exam questions based on ASME V.

T-676: interpretation of PT results

T-676.1: final interpretation

Final interpretation of the PT results has to be made within 10 to 60 minutes after the developer has dried. If bleed-out does not alter the examination results, longer periods are permitted. If the surface to be examined is too large to complete the examination within the prescribed or established time, the examination should be performed in increments.

This is simply saying: *inspect within 10–60 minutes*. A longer time can be used if you expect very fine imperfections. Very large surfaces can be split into sections.

T-676.2: characterizing indication(s)

Deciding (called *characterizing* in ASME-speak) the types of discontinuities can be difficult if the penetrant diffuses excessively into the developer. If this condition occurs, close observation of the formation of indications during application of the developer may assist in characterizing and determining the extent of the indications; i.e. the shape of deep indications can be masked by heavy leaching out of the penetrant, so it is advisable to start the examination of the part as soon as the developer is applied.

T-676.4: fluorescent penetrants

With fluorescent penetrants, the process is essentially the same as for colour contrast, but the examination is performed using an ultraviolet light, sometimes called *black light*. This is performed as follows:

(a) It is performed in a darkened area.
(b) The examiner must be in the darkened area for at least 5 minutes prior to performing the examination to enable his or her eyes to adapt to dark viewing. He or she must not wear photosensitive glasses or lenses.
(c) Warm up the black light for a minimum of 5 min prior to use and measure the intensity of the ultraviolet light emitted. Check that the filters and reflectors are clean and undamaged.
(d) Measure the black light intensity with a black lightmeter. A minimum of 1000 $\mu W/cm^2$ on the surface of the part being examined is required. The black light intensity must be re-verified at least once every 8 hours, whenever the workstation is changed or whenever the bulb is changed.

T-680: evaluation of PT indications

Indications are evaluated using the relevant code acceptance criteria (e.g. B31.3 for pipework). Remember that ASME V does not give acceptance criteria. Be aware that false indications may be caused by localized surface irregularities. Broad areas of fluorescence or pigmentation can mask defects and must be cleaned and re-examined.

Now try these familiarization questions on ASME V articles 1, 2 and 6.

16.5 ASME V articles 1, 2 and 6: familiarization questions

Q1. ASME section V article 2: radiography T-223
When performing a radiograph, where is the 'backscatter indicator' lead letter 'B' placed?

(a) On the front of the film holder ☐
(b) On the outside surface of the pipe ☐
(c) On the internal surface of the pipe ☐
(d) On the back of the film holder ☐

Q2. ASME section V article 2: radiography T-277.1 (d)
Wire IQIs must be placed so that they are:

(a) At 45° to the weld length ☐
(b) Parallel to the weld metal's length ☐
(c) Perpendicular to the weld metal's longitudinal axis but not across the weld ☐
(d) Perpendicular to the weld metal's longitudinal axis and across the weld ☐

Q3. ASME section V article 6: penetrant testing T-620
Liquid penetrant testing can be used to detect:

(a) Subsurface laminations ☐
(b) Internal flaws ☐
(c) Surface and slightly subsurface discontinuities ☐
(d) Surface breaking discontinuities ☐

Q4. ASME section V article 1: T-150
When an examination to the requirements of section V is required by a code such as ASME B31.3 the responsibility for establishing NDE procedures lies with:

(a) The inspector ☐
(b) The examiner ☐
(c) The user's quality department ☐
(d) The installer, fabricator or manufacturer/repairer ☐

Q5. ASME section V article 6: penetrant testing mandatory appendix II
Which penetrant materials must be checked for the following contaminants when used on austenitic stainless steels?

(a) Chlorine and sulphur content ☐

(b) Fluorine and sulphur content ☐
(c) Fluorine and chlorine content ☐
(d) Fluorine, chlorine and sulphur content ☐

16.6 ASME V article 7: magnetic testing (MT)

Similar to the previous article 6 covering penetrant testing, this article 7 of ASME V explains the technical principle of magnetic testing (MT). As with PT, we have already covered much of this in API 577, but article 7 adds more formal detail. Remember again that it is not component specific; it deals with the MT techniques themselves, not the *extent* of MT you have to do on a pressure vessel.

T-720: general
MT methods are used to detect cracks and other discontinuities on or near the surfaces of ferromagnetic materials. It involves magnetizing an area to be examined, then applying ferromagnetic particles to the surface, where they form patterns where the cracks and other discontinuities cause distortions in the normal magnetic field.

Maximum sensitivity is achieved when linear discontinuities are orientated *perpendicular to the lines of magnetic flux*. For optimum effectiveness in detecting all types of discontinuities, each area should therefore be examined at least *twice*, with the lines of flux during one examination approximately perpendicular to the lines of flux during the other; i.e. you need two field directions to do the test properly.

T-750: the MT techniques (see Fig. 16.5)
One or more of the following five magnetization techniques can be used:

(a) Prod technique
(b) Longitudinal magnetization technique
(c) Circular magnetization technique
(d) Yoke technique
(e) Multidirectional magnetization technique

The NDE Requirements of ASME V

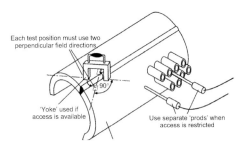

MT prod and yoke methods

Figure 16.5   MT examination technique

The API examination will be based on the prod or yoke techniques (i.e. (a) or (d) above), so these are the only ones we will consider. The others can be ignored for exam purposes.

T-752: the MT prod technique

T-752.1: the magnetizing procedure
Magnetization is accomplished by pressing portable prod type electrical contacts against the surface in the area to be examined. To avoid arcing, a remote control switch, which may be built into the prod handles, must be provided to allow the current to be turned on *after* the prods have been properly positioned.

T-752.3: prod spacing
Prod spacing must not exceed 8 in (203 mm). Shorter spacing may be used to accommodate the geometric limitations of the area being examined or to increase the sensitivity, but prod spacings of less than 3 in (76 mm) are usually not practical due to 'banding' of the magnetic particles around the prods. The prod tips must be kept clean and dressed (to give good contact).

T-755: the MT yoke technique
This method must only be used (either with AC or DC electromagnetic yokes or permanent magnet yokes) to detect discontinuities that are *surface breaking* on the component.

T-764.1: magnetic field strength
When doing an MT test, the applied magnetic field must have sufficient strength to produce satisfactory indications, but it must not be so strong that it causes the masking of relevant indications by non-relevant accumulations of magnetic particles. Factors that influence the required field strength include:

- Size, shape and material permeability of the part
- The magnetization technique
- Coatings
- The method of particle application
- The type and location of discontinuities to be detected

Magnetic field strength can be verified by using one or more of the following three methods:

- Method 1: T-764.1.1: pie-shaped magnetic particle field indicator
- Method 2: T-764.1.2: artificial flaw shims
- Method 3: T-764.1.3 hall effect tangential-field probe

T-773: methods of MT examination (dry and wet)
Remember the different types of MT technique. The ferromagnetic particles used as an examination medium can be either *wet* or *dry*, and may be either fluorescent or colour contrast:

- For dry particles the magnetizing current *remains on* while the examination medium is being applied and excess of the examination medium is removed. Remove the excess particles with a light air stream from a bulb, syringe or air hose (see T-776).
- For wet particles the magnetizing current will be *turned on*

after applying the particles. Wet particles from aerosol spray cans may be applied before and/or after magnetization. Wet particles can be applied during magnetisation as long as they are not applied with sufficient velocity to dislodge accumulated particles.

T-780: evaluation of defects found during MT
As with the other NDE techniques described in ASME V, defects and indications are evaluated using the relevant code acceptance criteria (e.g. ASME B31.3). Be aware that false indications may be caused by localized surface irregularities. Broad areas of particle accumulation can mask relevant indications and must be cleaned and re-examined.

16.7 ASME V article 23: ultrasonic thickness checking

In the ASME V code, this goes by the grand title of *Standard Practice for Measuring Thickness by Manual Ultrasonic Pulse-Echo Contact Method: section SE-797.2*. This makes it sound much more complicated than it actually is. Strangely, it contains some quite detailed technical requirements comprising approximately seven pages of text and diagrams at a level that would be appropriate to a UT qualification exam. The underlying principles, however, remain fairly straightforward. We will look at these as broadly as we can, with the objective of picking out the major points that may appear as closed-book questions in the API examinations.

The scope of article 23, section SE-797

This technique is for measuring the thickness of any material in which ultrasonic waves will propagate at a constant velocity and from which back reflections can be obtained and resolved. It utilizes the *contact pulse echo method* at a material temperature not to exceed 200 °F (93 °C). Measurements are made from one side of the object, without requiring access to the rear surface.

The idea is that you measure the velocity of sound in the

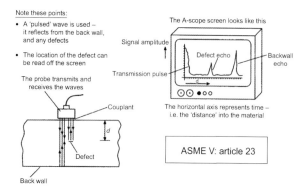

It simply uses the time it takes a pulsed sound wave to pass through a material to give a measure of its thickness

Figure 16.6 UT thickness checking

material and the time taken for the ultrasonic pulse to reach the back wall and return (see Fig. 16.6). Halving the result gives the thickness of the material.

Summary of practice

Material thickness (T), when measured by the pulse-echo ultrasonic method, is a product of the velocity of sound in the material and one half the transit time (round trip) through the material. The simple formula is:

$$T = Vt/2$$

where

T = thickness
V = velocity
t = transit time

Thickness-checking equipment

Thickness-measurement instruments are divided into three groups:

Flaw detectors with CRT readouts. These display time/amplitude information in an *A-scan* presentation (we saw this method in a previous module). Thickness is measured by reading the distance between the zero-corrected initial pulse and first-returned echo (back reflection), or between multiple-back reflection echoes, on a calibrated base-line of a CRT. The base-line of the CRT should be adjusted to read the desired thickness increments.

Flaw detectors with CRT and direct thickness readout. These are a combination pulse ultrasound flaw detection instrument with a CRT and additional circuitry that provides digital thickness information. The material thickness can be electronically measured and presented on a digital readout. The CRT provides a check on the validity of the electronic measurement by revealing measurement variables, such as internal discontinuities, or echo-strength variations, which might result in inaccurate readings.

Direct thickness readout meters. Thickness readout instruments are modified versions of the pulse-echo instrument. The elapsed time between the initial pulse and the first echo or between multiple echoes is converted into a meter or digital readout. The instruments are designed for measurement and direct numerical readout of specific ranges of thickness and materials.

Standardization blocks

Article 23 goes into great detail about different types of 'search units'. Much of this is too complicated to warrant too much attention. Note the following important points.

Section 7.2.2.1: calibration (or standardization) blocks
Two 'calibration' blocks should be used: one approximately the *maximum* thickness that the thickness meter will be measuring and the other the *minimum* thickness.

Thicknesses of materials at high temperatures up to about 540 °C (1000 °F) can be measured with specially designed instruments with high-temperature compensation. A rule of thumb is as follows:

- A thickness meter reads 1 % too high for every 55 °C (100 °F) above the temperature at which it was calibrated. This correction is an average one for many types of steel. Other corrections would have to be determined empirically for other materials.
- **An example**. If a thickness meter was calibrated on a piece of similar material at 20 °C (68 °F), and if the reading was obtained with a surface temperature of 460 °C (860 °F), the apparent reading should be reduced by 8 %.

Now try these familiarization questions covering ASME V articles 7 and 23 (article 9 questions are too easy).

16.8 ASME V articles 7 and 23 familiarization questions

Q1. ASME section V article 7: magnetic particle testing T-720
Magnetic particle testing can be used to find:

(a) Surface and near-surface discontinuities in all materials ☐
(b) Surface and near-surface discontinuities in ferromagnetic materials ☐
(c) Surface and near-surface discontinuities in all metallic materials ☐
(d) Surface breaking discontinuities only ☐

Q2. ASME section V article 7: magnetic particle testing T-720
During an MT procedure, maximum sensitivity for finding discontinuities will be achieved if:

(a) The lines of magnetic flux are perpendicular to a linear discontinuity ☐
(b) The lines of magnetic flux are perpendicular to a volumetric discontinuity ☐
(c) The lines of magnetic flux are parallel to a linear discontinuity ☐
(d) The lines of magnetic flux are parallel to a volumetric discontinuity ☐

Q3. ASME section V article 7: magnetic particle testing T-741.1 (b)

Surfaces must be cleaned of all extraneous matter prior to magnetic testing. How far back must adjacent surfaces to the area of interest be cleaned?

(a) At least 2 inches ☐
(b) At least $\frac{1}{2}$ inch ☐
(c) Cleaning is not required on adjacent surfaces ☐
(d) At least 1 inch ☐

Q4. ASME section V article 7: magnetic particle testing T-741.1 (d)

According to ASME V, what is the maximum coating thickness permitted on an area to be examined by MT?

(a) 50 µm ☐
(b) No coating is permitted ☐
(c) 40 µm ☐
(d) An actual value is not specified ☐

Q5. ASME section V article 7: magnetic particle testing T-764.1

Which of the following methods can verify the adequacy of magnetic field strength?

(a) A pie-shaped magnetic particle field indicator ☐
(b) Artificial flaw shims ☐
(c) A gaussmeter and Hall effect tangential-field probe ☐
(d) They can all be used ☐

Q6. ASME section V article 7: magnetic particle testing T-762(c)

What is the lifting power required of a DC electromagnet or permanent magnet yoke?

(a) 40 lb at the maximum pole spacing that will be used ☐
(b) 40 lb at the minimum pole spacing that will be used ☐
(c) 18.1 lb at the maximum pole spacing that will be used ☐
(d) 18.1 lb at the minimum pole spacing that will be used ☐

Q7. ASME section V article 7: magnetic particle testing T-752.2

Which types of magnetizing current can be used with the prod technique?

(a) AC or DC ☐
(b) DC or rectified ☐
(c) DC only ☐
(d) They can all be used ☐

Q8. ASME section V article 7: magnetic particle testing T-752.3

What is the maximum prod spacing permitted by ASME V?

(a) It depends on the current being used.. ☐
(b) There is no maximum specified in ASME codes ☐
(c) 8 inches ☐
(d) 6 inches ☐

Q9. ASME section V article 7: magnetic particle testing T-755.1

What is the best description of the limitations of yoke techniques?

(a) They must only be used for detecting surface breaking discontinuities ☐
(b) They can also be used for detecting subsurface discontinuities ☐
(c) Only AC electromagnet yokes will detect subsurface discontinuities ☐
(d) They will detect linear defects in austenitic stainless steels ☐

Q10. ASME section V article 7: magnetic particle testing, appendix 1

Which MT technique is specified in ASME article 7 mandatory appendix 1 to be used to test coated ferritic materials?

(a) AC electromagnet ☐
(b) DC electromagnet ☐
(c) Permanent magnet ☐
(d) AC or DC prods ☐

Q11. ASME section V article 23: ultrasonic thickness testing, section 5.1

UT thickness checking using standard equipment is used for temperatures up to:

(a) 93 °F ☐
(b) 200 °F ☐
(c) 150 °F ☐
(d) 150 °C ☐

Q12. ASME section V article 23: ultrasonic thickness testing, section 8.5

Special ultrasonic thickness measurement equipment can be used at high temperatures. If the equipment is calibrated at ambient temperature, the apparent thickness reading displayed at an elevated temperature should be:

(a) Reduced by 1 % per 55 °F ☐
(b) Increased by 1 % per 55 °F ☐
(c) Increased by 1 % per 100 °F ☐
(d) Reduced by 1 % per 100 °F ☐

Chapter 17

Thirty Open-Book Sample Questions

Try these questions, using all of the codes specified in the API 510 exam 'effectivity list'. There is little point in guessing the answers – the objective is to see where the answers come from in the codes, thereby increasing your familiarity with the content.

Question 1
A vessel is constructed to a pre-1999 version of the ASME construction code. Can it be re-rated to the 1999 version?

(a) No, under no circumstances does API 510 allow it ☐
(b) Yes, if it is re-tested to 130 % MAWP corrected for temperature ☐
(c) Yes, if it is permitted by API 510 fig. 8-1 ☐
(d) Yes, without restriction ☐

Question 2
The API philosophy is that temporary repairs should be replaced with permanent repairs:

(a) Within 12 months ☐
(b) At the next available maintenance opportunity ☐
(c) The next time the plant is shut down ☐
(d) At the next internal inspection ☐

Question 3
A procedure is qualified using base material with an S-number. Which of the following statements is true?

a) This qualifies corresponding S-number materials only ☐
b) This qualifies corresponding F-number materials ☐
c) This qualifies corresponding P-number materials only ☐
d) This qualifies both P-number and S-number materials ☐

Question 4
Which of the following NDE methods would be *unlikely* to find an edge breaking lamination in a weld joint?

(a) MT ☐
(b) PT ☐
(c) RT ☐

Thirty Open-Book Sample Questions

(d) UT ☐

Question 5
The inspection interval for all PRVs is determined by:

(a) The inspector ☐
(b) The engineer ☐
(c) Any other qualified individual listed in the owner/user's QA system ☐
(d) Any of the above ☐

Question 6
Who decides the corrosion rate that best reflects current process conditions?

(a) The inspector ☐
(b) The corrosion specialist ☐
(c) The inspector in consultation with the corrosion specialist ☐
(d) The owner/user (in the written scheme of examination) ☐

Question 7
Providing certain conditions are met, a vessel may be re-rated to:

(a) Its original construction code only ☐
(b) Its original construction code and the later edition of the construction code ☐
(c) The latest edition of the construction code only ☐
(d) The ASME code only ☐

Question 8
Thickness data for a pressure vessel are provided as follows:

Minimum (calculated) thickness = 0.125 in
Current measured thickness = 0.25 in
Measured thickness 4 years ago = 0.50 in

What is the remaining life and the inspection period of the vessel?

(a) 4 years, 2 years ☐
(b) 4 years, 4 years ☐
(c) 2 years, 2 years ☐
(d) None of the above ☐

Question 9
Weld repairs to existing stainless steel overlay or clad areas should consider:

(a) The problem of local softening ☐
(b) The problem of martensite formation ☐
(c) The problem of temper embrittlement ☐
(d) The possibility of increased corrosion rates after the repair is completed ☐

Question 10
A pressure vessel has the following data:

- Nameplate stamping RT-2
- MAWP = 280 psig at 690 °F
- S = 13 350 psi
- Actual shell thickness = 0.475 in
- ID = 78.5 in
- Welded joints are all type 1

What is the minimum safe shell thickness to support the rated MAWP of 280 psi?

(a) Approximately 0.83 in ☐
(b) Approximately 1.10 in ☐
(c) Approximately 0.75 in ☐
(d) Approximately 0.89 in ☐

Question 11
For vessels susceptible to brittle fracture, what additional tests should the inspector specifically consider after repair welding is completed?

(a) Additional volumetric NDE ☐
(b) Additional surface NDE ☐
(c) Metallographic tests ☐
(d) Shear wave (angle probe) UT ☐

Question 12
A vessel has the following data:

- Elliptical heads joined to the shell by type 1 joints (Cat B) with full RT
- Head ID = 52 in
- Thickness at the knuckle (corroded) = 0.28 in

- $S = 13\,800$ psi

What is the safe MAWP of this head (ignoring any static pressure consideration)?

a) 137 psi
b) 148 psi
c) 175 psi
d) 190 psi

Question 13
For a vessel that has no nameplate, the inspector should specify that a new nameplate be fitted showing:

(a) Maximum and minimum allowable temperature, MAWP and date
(b) MAWP, MDMT and construction code
(c) MAWP, maximum and minimum allowable temperature, and API 510 stamp
(d) All of the above plus the ASME U-stamp

Question 14
RBI assessment should be thoroughly documented in accordance with:

(a) API 580 section 10
(b) API 580 section 16
(c) API 510 section 6
(d) All of the above

Question 15
In API 510, sulphidation is classed as a DM resulting in:

(a) General/local metal loss
(b) Surface connected cracking
(c) Metallurgical changes
(d) Blistering

Question 16
Which of these is not a material verification (PMI) technique that can be used on site?

(a) Spectrographic analysis
(b) Metallographic replication
(c) XRF
(d) Spark testing

Question 17
Plain carbon and other ferritic steels may be in danger of brittle fracture at:

a) Above 380 °C ☐
b) 60–120 °C ☐
c) Ambient temperature ☐
d) None of the above ☐

Question 18
A vessel constructed of material with a thickness of 0.50 in and UTS of 75 000 psi is to be weld-repaired using weld consumables with UTS of 60 000 psi. If the depth of the weld repair is 0.2 in, what is the total required thickness of the weld deposit?

(a) 0.16 in ☐
(b) 0.20 in ☐
(c) 0.250 in ☐
(d) 0.275 in ☐

Question 19
A vessel is made from carbon steel plate (UCS-66 curve A) with a design stress of 17 500 psi. It operates at very low material stress of 100 psi and is made of 1 in plate that has been spot radiographed. What is the minimum design metal temperature for the material to not require impact testing?

(a) −100 °F ☐
(b) −25 °F ☐
(c) −72 °F ☐
(d) −55 °F ☐

Question 20
It is required to investigate whether a scheduled internal inspection on a multizone vessel (with varying corrosion rates) can be substituted by an on-stream inspection. The inspector should:

(a) Specify an RBI assessment ☐
(b) Take the worst-corroding zone as the reference ☐
(c) Treat each zone independently ☐
(d) Prohibit the substitution as it is not allowed for multizone vessels according to API 510 ☐

Thirty Open-Book Sample Questions

Question 21
If there is conflict between the ASME codes and API 510 then:

(a) ASME takes priority ☐
(b) API 510 takes priority ☐
(c) Both codes shall be considered equally ☐
(d) The owner/user shall decide which takes priority ☐

Question 22
A nozzle is fitted abutting (i.e. *set-on*) the vessel wall. What is an acceptable method of attaching it?

(a) With a full penetration groove weld through the nozzle wall ☐
(b) With a full penetration groove weld through the vessel wall ☐
(c) With a partial penetration groove weld through the nozzle wall ☐
(d) Both (a) and (c) are acceptable ☐

Question 23
An engineer has passed the NBBPVI inspection examination. Under what conditions can the inspector be awarded an API 510 authorized pressure vessel inspector certificate?

(a) By just proving that he or she meets the API 510 education and experience requirements ☐
(b) By demonstrating a further 1 year's experience ☐
(c) By applying to NBBPVI for a concession ☐
(d) Only by passing the API 510 exam ☐

Question 24
A vessel has an inside diameter of 30 inches. What is the maximum allowed averaging length for calculating corroded wall thickness?

(a) 10 in ☐
(b) 15 in ☐
(c) 20 in ☐
(d) 30 in ☐

Question 25
What are HIC cracks likely to look like?

(a) A spider's web ☐
(b) A set of stairs ☐

(c) A series of roughly shaped hexagons ☐
(d) Daggers ☐

Question 26
Which of the following is a commonly accepted advantage of the GMAW process?

(a) It has a high deposition rate compared to SMAW or GTAW ☐
(b) It has the best control of the weld pool of any of the arc processes ☐
(c) It is less sensitive to wind and draughts than any other process ☐
(d) It gives deep penetration on thick steels in a short-circuiting transfer ☐

Question 27
How often should an external inspection be performed on an above-ground vessel?

(a) 5 years ☐
(b) 10 years ☐
(c) Halfway through the calculated remaining life ☐
(d) It depends on the process ☐

Question 28
Which of these vessels in which internal inspection is physically possible, but are in severe corrosive service, may not use an on-stream inspection as a substitute for an internal inspection?

(a) The recorded corrosion rate is 0.1 mm/year ☐
(b) The vessel is 3 years old ☐
(c) The vessel operates at 80 °C ☐
(d) The vessel has no protective internal lining ☐

Question 29
For a vessel that has no nameplate or design/construction information, a pressure test should be performed:

(a) Immediately ☐
(b) Before any further use of the vessel ☐
(c) At the next scheduled inspection ☐
(d) As soon as practical ☐

Question 30
CMLs should be distributed:

(a) Appropriately over a vessel ☐
(b) In highly stressed areas ☐
(c) In areas of proven corrosion ☐
(d) Near areas of past failure ☐

Chapter 18

Answers

18.1 Familiarization answers

Subject question and Chapter	Question number	Answer
API 510 (sections 1-4) (Chapter 2)	1 2 3 4 5 6 7 8 9 10	a d c c c a d b a b
API 510 (section 5): inspection practices (Chapter 3)	1 2 3 4 5 6 7 8 9 10 11 12	b a b c b c a b b a c b
API 510 (section 6): inspection periods (Chapter 4)	1 2 3 4 5 6 7 8	a a c a b b b c
(See next page)	1 2 3	a b a

292

Answers

Subject question and Chapter	Question number	Answer
API 510 (section 7): inspection data evaluation (Chapter 4)	4 5 6 7 8 9 10	b b a d c c c
API 510 (section 8): repair, alteration, re-rating (Chapter 5)	1 2 3 4 5 6 7 8 9 10	a d c c a b c c a b
API 572: inspection of vessels (Chapter 6)	1 2 3 4 5 6 7 8 9 10	c d b c a d d c b a
API 571 (set 1): damage mechanisms (Chapter 7)	1 2 3 4 5 6 7 8 9 10	b c c a c b c b b b
API 571 (set 2): damage mechanisms (Chapter 7)	1 2 3	b b a

Quick Guide to API 510

Subject question and Chapter	Question number	Answer
	4	b
	5	d
API 576: PRVs (Chapter 8)	1	b
	2	b
	3	d
	4	a
	5	d
	6	a
	7	b
	8	d
	9	d
	10	c
	11	c
	12	a
	13	b
	14	c
	15	d
ASME VIII: pressure design (set 1): internal pressure (Chapter 9)	1	d
	2	a
	3	c
	4	a
	5	c
	6	d
	7	c
	8	a
	9	c
	10	b
ASME VIII: pressure design (set 2): MAWP and pressure testing (Chapter 9)	1	a
	2	b
	3	c
	4	d
	5	b
	6	b
	7	c
	8	c
	9	b
	10	c
ASME VIII: Pressure design (set 3):	1	d
	2	c
	3	d

Answers

Subject question and Chapter	Question number	Answer
external pressure (Chapter 9)	4 5	c a
ASME VIII (set 1: UW-11): welding and NDE (Chapter 10)	1 2 3 4 5	d c c a a
ASME VIII (set 2: UW-16): welding and NDE (Chapter 10)	1 2 3 4 5	a d c a a
ASME VIII (set 3: UW-51/52): welding and NDE (Chapter 10)	1 2 3 4 5	b c d b d
ASME VIII UCS-56/UW-40: PWHT (Chapter 11)	1 2 3 4 5	b d b d b
ASME VIII UCS-66: impact test exemption (Chapter 12)	1 2 3	a b b
API 577: welding process (Chapter 13)	1 2 3 4 5 6 7 8 9 10	a b b a d d c c b b
(See next page)	1 2 3	a b b

295

Quick Guide to API 510

Subject question and Chapter	Question number	Answer
API 577 welding consumables (Chapter 13)	4	a
	5	a
	6	d
	7	c
	8	c
	9	b
	10	d
ASME IX articles I and II (Chapter 14)	1	c
	2	b
	3	d
	4	a
	5	c
	6	a
	7	b
	8	b
	9	d
	10	c
ASME IX articles III and IV (Chapter 14)	1	d
	2	d
	3	d
	4	b
	5	a
	6	a
	7	a
	8	a
	9	c
	10	d
ASME V articles 1, 2 and 6 (Chapter 16)	1	d
	2	d
	3	d
	4	d
	5	c
ASME V Articles 7 and 23 (Chapter 16)	1	b
	2	a
	3	d
	4	d
	5	d
	6	a
	7	b
	8	c

Answers

Subject question and Chapter	Question number	Answer
	9	a
	10	a
	11	b
	12	d

18.2 Open-book sample questions answers

Question 1. ANS c
API 510 section 8.2 (b): Re-rating

Question 2. ANS b
API 510 section 8.1.5.1: defect repairs (temporary)

Question 3. ANS a
ASME IX section QW-420.2

Question 4. ANS c
API 577 Section 9 table 5

Question 5. ANS d
API 510 section 6.6.2.1: PRV inspection intervals

Question 6. ANS c
API 510 section 7.1.1.2: corrosion rate determination

Question 7. ANS b
API 510 section 8.2.1: (b) Re-rating

Question 8. ANS c
API 510 corrosion rate calculation

Metal loss $= 0.500 \text{ in} - 0.250 \text{ in} = 0.250 \text{ in}$

Corrosion rate $= \dfrac{0.250 \text{ in}}{4} = 0.0625$ per year

Corrosion allowance $= 0.125 \text{ in}$

Remaining life $= \dfrac{0.125 \text{ in}}{0.0625} = 2$ years

Inspection interval $= 2$ years (from API 510)

Question 9. ANS c
API 510 section 8.1.5.4.2: repairs to stainless steel overlay and cladding

Question 10. ANS a
UG-27: shell thickness with RT-2 stamping
  Applying shell formula UG-27
  Minimum shell thickness $t = PR/(SE–0.6P)$
  Determine E; RT-2 marking indicates that UW-11 (a)(5)(b) has been complied with so we can use $E = 1$ (also table UW-12 for type 1 weld). The 'full RT' interpretation of RT-2 gives $E = 1$, where

$S = 13\,350$ psi
$E = 1.0$ (from table UW-12)
$R = 39.25$ in (1/2 ID of 78.5 in)
$P = 280$ psig
and

$$t = \frac{280 \times 39.25}{(13\,350 \times 1.0) - (0.6 \times 280)} = \frac{10\,990}{13\,350 - 168} = 0.833 \text{in}$$

Question 11. ANS b
API 510 section 8.1.8: weld inspection for vessels subject to brittle fracture

Question 12. ANS b
ASME VIII UG-32: head thickness with E from UW-12 for full RT
  Simple application of UG-32
  Determine E from UW-12
  Cat B, type 1 full RT gives $E = 1$
  Pressure: $P = $ elliptical head
  Given: $S = 13\,800$
  $E = 1.0$
  $D = 52$ in
  $t = 0.28$ in

$$P = \frac{2 \times 13\,800 \times 0.28}{52 + (0.2 \times 0.28)} = \frac{7728}{52 + 0.056} = 148.45 \text{ psi}$$

$P = 148.45$ psi   ANS

Question 13. ANS a
API 510 section 7.7: equipment with minimal documentation

Answers

Question 14. ANS b
API 510 section 5.2.3: RBI documentation

Question 15. ANS a
API 510 section 5.4: damage modes

Question 16. ANS b
API 510 section 5.9.1: material verification

Question 17. ANS c
API 510 section 5.8.6: pneumatic test temperature

Question 18. ANS c
API 510 section 8: weld repairs using different materials
 API 510, section 8.1.5.3.2: specifies that the thickness of a repair weld has to be increased by the ratio of the specified minimum UTS of the repair metal and the base metal:

$$\frac{\text{Tensile strength of base metal}}{\text{Tensile strength of weld metal}} = \frac{75}{60} = 1.25$$

Required total thickness of weld deposit $= 0.2 \times 1.25$
$= 0.25$ in Ans

Question 19. ANS d
ASME VIII table UCS-66
 Take figure UCS-66M. For 1 in (25 mm) with curve A minimum design temperature is 68 °F. For stress ratio less than 0.35 reduction in MDMT can be 140 °F (far left of curve). MDMT could be 68–140 = –72°F, but this is limited by UCS-66 (b)(1)(a) to –55°F.

Question 20. ANS c
API 510 section 6.5.3 multizone vessels

Question 21. ANS b
API 510 section 1.1.1

Question 22. ANS a
ASME VIII section UW-16 (c) and sketches (a), (b). Necks attached by a full penetration weld

Question 23. ANS a
API 510 appendix B.2.2: inspector certification

Question 24. ANS b
API 510 section 7.4.2: evaluation of locally thinned areas

Question 25. ANS b
API 571 section 5.1.2.3.1: wet H_2S damage types

Question 26. ANS a
API 577 section 5.4.4

Question 27. ANS a
API 510 section 6.4.1: external inspection period of 5 years specified

Question 28. ANS b
API 510 section 6.5.2.1 (b)(3): on-stream inspections. Corrosive character of the contents must be established over a minimum of 5 years.

Question 29. ANS d
API 510 section 7.7: equipment with minimal documentation

Question 30. ANS a
API 510 section 5.6.3.1: CML selection: exact wording of code

Chapter 19

The Final Word on Exam Questions

19.1 All exams anywhere: some statistical nuts and bolts for non-mathematicians

The statistiscs of exam questions are unlikely to tax the brains of eminent statisticians for very long. Let's say we want to choose 150 questions randomly from a set of, say, 900 if that's how many we have. Now take one question, the classic query about, among other things, how many angels can realistically be persuaded to dance on the head of a pin. Let's call this, for convenience, question P (P for pin). Every so often, when choosing the question set for our exam, question P will no doubt appear; but how often?

Take the first exam cycle; if we choose 150 questions truly randomly from one large set of 900, then the chance of our question P appearing in the exam is precisely 150/900 or 1 in 6 (or 16.67 % if you like). Put another way, over time it will appear once in every six exams. This means, of course, that if you keep on taking the exam again and again, five times out of every six any effort you have put into remembering the answer to question P (the answer is 14 incidentally) will have been totally wasted.

But hold on; the situation has changed. You now know the answer, so *we want it* to appear next time. At the next exam cycle, the chances of P appearing are once again 16.67 (but the cumulative probability of the two successive appearances, given that it already had only a 16.67 % of appearing last time) are much less ... let's say 1 in 36. Extending this out, the probability of P turning up in three successive exam cycles are getting pretty thin and in four successive cycles, miniscule at best. The odds against you are awful ... it's looking like your valuable knowledge of the P answer will most of the time be wasted.

Just when you thought things were looking bad ... it gets

worse. Waiting in the wings is another question Q, to which you already know the answer (it's the one about the length of the piece of string). It would be nice if this appeared regularly, so you could confidently tick the correct answer box marked 17 inches. Even better, what if both P and Q appeared in every exam draw now you could get them both right. What's the chance of this?

- The probability of P and Q both appearing in the first draw are small – we know that
- In the first and second draw, an order of magnitude smaller
- In the first, second and third draws .... undeniably tiny
- And, in the first, second, third and fourth draws ... miniscule would be the operative word

Extending this out to say, five questions, P, Q, R, S and T, that you are certain of the answer, the chances of all five appearing in three successive draws are so near zero that the calculation would be enough to leave your calculator a blackened ruin.

But wait ... if, by some stretch of the imagination, such a thing *did* actually happen, you would have to conclude that either these lottery-type odds had occurred or that there was maybe some other explanation. But what could it be?

Working the maths backwards would tell us that we could only be certain to get such a run of unlikely probabilities if we had not 900 questions to choose from at all but a significantly smaller set. This is the much-reduced set size we would need if we were still drawing 150 purely at random from one big set and our five questions P, Q, R, S and T all miraculously appeared in each of three successive draws. That's one possibility.

There's another way to do it. If we divide our bank of 900 questions into, say, 10 sets of 90 questions each, based on subject breakdown, and let's say our exam draw will require that we draw 15 questions from each set to give us the 150-question draw. If our favourite questions P, Q, R, S and T each reside in a different set (which they will, as they are about different subjects), then the odds of all five appearing in three successive draws are reasonably believable, with a little imagination at least. Adding some regular preferences *within* each set would reduce the odds even further.

In the final act, with a little biasing towards certain questions in each small set our question-drawing exercise will throw off its mask of randomness and, right on cue, up will pop our P, Q, R, S, T combination with a flourish, like the demon king among a crowd of pantomime fairies.

So if you ever see this, the explanation may be one of the options above.

19.2 Exam questions and the three principles of whatever (the universal conundrum of randomness versus balance)

As with most engineering laws and axioms (pretend laws) you won't get far without a handful of principles (of whatever).

The first principle (of whatever) is that, faced with the dilemma between randomness and balance, any set of exam questions is destined to end up with a bit of both. A core of balance (good for the technical reputation of the whole affair) will inevitably be surrounded by a shroud of some randomness, to pacify the technically curious, surprise the complacent and frustrate the intolerant – in more or less equal measure. There is nothing wrong with this; the purpose of any exam programme must be to weed out those candidates who are not good enough to pass.

Now we have started, the first principle spawns, in true Newtonian fashion, the second principle – a strategy for dealing with the self-created problems of the first. The problem is the age-old one of *high complexity*. Code documents contain tens of thousands of technical facts, each multifaceted, and together capable of being assembled into an almost infinite set of exam questions. We need some way to deal with this. The second principle becomes: *selectivity can handle this complexity*.

Tightening this down, we get the third principle: *only selectivity can handle this complexity*. There's nothing academic about the third principle (of whatever); it just says that if you try to memorize and regurgitate, brightly coloured parrot-fashion, *all* the content of any exam syllabus, you are almost guaranteed to fail. You will fail because most of the time the high complexity will get you. It has to, because exam questions can replicate and mutate in almost infinite variety, whereas you cannot. You may be lucky (who doesn't need a bit of luck?) but a more probable outcome is that you will be left taking the exam multiple times. Round and round and round you will go at your own expense, clawing at the pass/fail interface.

A quick revisit of the first principle (of whatever) suggests that being selective in the parts of an exam syllabus we study carries with it a certain risk. The price for being selective is that you may be wrong. Most of the risk has its roots in the amount of balance versus randomness that exists in the exam set. The more balanced it is, the more predictable it will be and the better your chances. Don't misread the situation though; your chances will never be any worse than they would have been if you hadn't been selective. The third principle tells us that.

Remembering this, you should only read the tables in section 19.3 if you subscribe to the three principles *and* you think selectivity is for you. If you don't recognize the code references, clause numbers or abbreviations then you need to start again at the beginning of this book.

19.3 Exam selectivity

For the wise	
Question	Subject: open book
1	RT density
2	RT backscatter symbol
3	Condensate corrosion
4	Hydrotest pressure
5	HTHA
6	RT joint type
7	Reformer failure
8	WFMT cleaning
9	CD welding
10	Wall thickness calculation
11	RT slag acceptance
12	Re-rating Fig. 8.1
13	Plate offset
14	NPS 2 nozzle to shell
15	PRV set pressure
16	MDMT
17	RT records
18	Vessel head calculation
19	Remaining life
20	Repair authorization
21	Dry MT temperatures
22	Corrosion averaging
23	PRV removal
24	P-number
25	Heat treatment
26	Average thickness
27	Charpy specimen length

28	Essential variables
29	CMLs
30	Temporary repair dimensions
31	Corrosion rate
32	pH values
33	Defects at weld toes
34	PWHT
35	Weld processes
36	API 579
37	Repair welding
38	Missing documents
39	CD welding (again)
40	Charpy values table
41	Pneumatic tests
42	Vessel linings
43	Corrosion buttons
44	Elliptical head calculation
45	RT step wedge
46	Sulphidation
47	CD welding in lieu of PWHT
48	Static head
49	Concrete foundations
50	Cooling water corrosion
For the hopeful*	
Question	**Subject: open book**
1	Factual questions from API 510 sections 1–4 that fit my experience
2	
3	
4	

The Final Word on Exam Questions

Question	Topic
5	Hard engineering logic questions from API 510 sections 5 and 6
6	
7	
8	Experience-based questions from API 510 section 7
9	
10	
11	API 510 section 8
12	
13	
14	
15	ASME VIII head and shell calculations (easy if you can use a calculator)
16	
17	
18	
19	Pressure testing questions (may need to consult the parrot)
20	
21	
22	
23	Easily found points from API 572 that are obvious to anyone in this inspection business
24	
25	
26	
27	
28	API 571 DM questions ... I'll have a guess at those ...
29	
30	
31	
32	
33	

34	NDE questions from ASME V ..... No problem with my previous experience. I used to be an NDE technician, you know
35	
36	
37	
38	
39	
40	
41	
42	
43	
44	Easily found points from API 577 that I agree with
45	
46	
47	ASME IX exercise (can be quite tricky ... hope they're not too hard)
48	
49	
50	

* Incorrect

Appendix

Publications Effectivity Sheet for API 570 Exam Administration: 2 June 2010

Listed below are the effective editions of the publications required for the **2 June 2010** API 510, Pressure Vessel Inspector Certification Examination.

- **API Standard 510,** *Pressure Vessel Inspection Code: In-Service Inspection, Rating, Repair, and Alteration*, 9th Edition, June 2006. IHS Product Code **API CERT 510**
- **API Recommended Practice 571,** *Damage Mechanisms Affecting Fixed Equipment in the Refining Industry*, 1st Edition, December 2003. IHS product code: **API CERT 510_571** (includes only the portions listed below)

ATTENTION: Only the following mechanisms to be included:

Par. 4.2.3 – Temper Embrittlement
    4.2.7 – Brittle Fracture
    4.2.9 – Thermal Fatigue
    4.2.14 – Erosion/Erosion Corrosion
    4.2.16 – Mechanical Failure
    4.3.2 – Atmospheric Corrosion
    4.3.3 – Corrosion Under Insulation (CUI)
    4.3.4 – Cooling Water Corrosion
    4.3.5 – Boiler Water Condensate Corrosion
    4.4.2 – Sulphidation
    4.5.1 – Chloride Stress Corrosion Cracking (ClSCC)
    4.5.2 – Corrosion Fatigue
    4.5.3 – Caustic Stress Corrosion Cracking (Caustic Embrittlement)
    5.1.2.3 – Wet H_2S Damage (Blistering/HIC/SOHIC/SCC)

5.1.3.1 – High Temperature Hydrogen Attack (HTHA)
- **API Recommended Practice 572**, *Inspection of Pressure Vessels*, 2nd Edition, February 2001. IHS Product Code **API CERT 572**
- **API Recommended Practice 576**, *Inspection of Pressure Relieving Devices*, 2nd Edition, December 2000. IHS Product Code **API CERT 576**
- **API Recommended Practice 577**, *Welding Inspection and Metallurgy*, 1st Edition, October 2004. IHS Product Code **API CERT 577**
- **American Society of Mechanical Engineers (ASME)**, *Boiler and Pressure Vessel Code*, 2007 Edition **with 2008 Addenda**
    i. **Section V**, *Nondestructive Examination, Articles 1, 2, 6, 7, 9 and 23 (section SE797 only)*
    ii. **Section VIII**, *Rules for Construction of Pressure Vessels, Division 1; Introduction (U), UG, UW, UCS, UHT, Appendices 14, 6, 8 and 12*
    iii. **Section IX**, *Welding and Brazing Qualifications, Welding Only*
    IHS Product Code for the ASME package **API CERT 510 ASME**. Package includes *only* the above excerpts necessary for the exam.

API and ASME publications may be ordered through IHS (formerly IHS Documents) at 00(1)-303-397-7956 or 00(1)-800-854-7179. Product codes are listed above. Orders may also be faxed to 00(1)-303-397-2740. More information is available at http://www.ihs.com. API members are eligible for a 30% discount on all API documents; exam candidates are eligible for a 20% discount on all API documents. When calling to order, please identify yourself as an exam candidate and/or API member. **Prices quoted will reflect the applicable discounts.** No discounts will be made for ASME documents.

Appendix

Note: API and ASME publications are copyrighted material. Photocopies of API and ASME publications are not permitted. CD-ROM versions of the API documents are issued quarterly by Information Handling Services and are allowed. Be sure to check your CD-ROM against the editions noted on this sheet.

Index

accumulation 109, 110
additional nameplate 68
advanced thinning analysis 57
alterations 13, 17, 62
  definition of 61
API 510
  Body of Knowledge 257
  contents 11
  NDE requirements 257–261
  NDE requirements, specific to 9th edition 261
  pressure–volume exemptions 12
  PWHT overrides 197–201
API 571 25
  Damage Mechanisms 89–104
API 572 75–88
API 576 105–123
  introduction to 105–106
API 579 51
  fitness-for-service evaluations 52–55
  sections of 55
API inspection codes 2
API recommended practice (RP) documents 6
API RP 577 Welding Inspection and Metallurgy 212
API RP 580 Risk-Based Inspection 24
approval of method 71
area compensation method 166
area replacement method 166
ASME construction codes 1
ASME IX
  article I 233–234
  article II 236–237
  article III 239–240
  article IV 240–243
  numbering system 233
  welding documentation 231
ASME V
  article 2: radiographic examination 263–269
  article 23: ultrasonic thickness checking 277–280
  article 6: penetrant testing (PT) 269–272
  article 7: magnetic testing (MT) 274–277
  NDE requirements 262–283
ASME VIII (UW-52) spot RT 190
ASME VIII
  clause numbering 128–129
  external pressure 126
  joint efficiency 126

post-weld heat treatment (PWHT) 194
Pressure Design 124–170
pressure testing 126
RT 186
RT grade 126, 127–128
weld categories 173
Welding and NDE 171–193
as-received pop pressure 122
atmospheric corrosion 98
authorization to proceed 71
authorized inspection agency 13
average remaining thickness 49
averaging area 49
averaging length 289
axial (longitudinal) stress 132

back pressure 110
backscatter 267, 268
bellows 116
blowdown 109
brittle fracture 28–29, 45, 92, 151, 203, 286, 288
buckling 156

caustic embrittlement 101
Charpy impact toughness test 203
code revisions 5
cold differential test pressure (CDTP) 110
composite rupture disc 115

condition monitoring locations (CMLs) 14, 27, 291
  selection 31
cone half angle 140
conical heads 140–141
contact pulse echo method 277
controlled deposition (CD) welding 198–201
corroded vessel heads 51, 54
corrosion
  allowance 51, 133, 141
  averaging 50
  fatigue 94
  mechanisms 80
  rate 45, 47, 104, 285
  rate determination 42–46
  under insulation (CUI) 97
CUI inspection 26

damage mechanisms (DMs) 25, 89–104
  atmospheric corrosion 98
  brittle fracture 92
  caustic embrittlement 101
  corrosion fatigue 94
  corrosion under insulation (CUI) 97
  erosion–corrosion 97
  high-temperature hydrogen attack (HTHA) 101

Index

mechanical fatigue 93
stress corrosion cracking (SCC) 100
sulphidation corrosion 99
temper embrittlement 103
thermal fatigue 92
wet H_2S damage 102
data evaluation 33
defect acceptance criteria 188
design of welded joints 179–181
design pressure 48, 144, 145
distance between stiffeners 157

effectivity list xii, xiii
ellipsoidal head 135–136
calculation example 136
design 143
essential variables 231, 241
evaluation
of locally thinned areas 49–50
of pitting 50–52
of vessels with minimal documents 58
exam questions 301–304
examination xiii
examiner 14
exemption for small nozzles 178
external inspection 26, 36, 83
external pressure 155

exam questions 157

failed spring 116, 121
fillet-weld tests 235
fitness for service (FFS) 1
fluorescent penetrants 272
flux-cored arc welding (FCAW) 215, 217
F-numbers 242
foundations and supports 86
full RT 186

geometric unsharpness, U_g 267
limitations 268
GMAW 290
guided-bend tests 235

half-life principle 43
half-life/double corrosion rate principle 44
hemispherical head geometry 139
HIC cracks 289
high-temperature hydrogen attack (HTHA) 101
hoop (circumferential) stress 130
hydrostatic head 46
hydrostatic test 146
procedure 29–30, 148–150

ICP (Individual Certification Program) xii

image quality indicators (IQIs) 263
  hole type 267
  placement 266
  sensitivity 267
  size to be used 264–265
impact exemption UCS-66 204–211
impact testing 203–211
imperfections and types of NDE method 262
incorrect materials 81
Individual Certification Program xii
inside crown radius 138
inspection data 42
inspection during installation 40
inspection during installation and service changes 35–36
inspection methods and limitations 82–84
inspection of pressure-relieving devices 105–123
inspection plans 23, 37
inspector
  recertification 21
  responsibilities 19
insulation removal 26
internal equipment 79
internal inspections 83–84
  interval 41

joint design 171–174
joint efficiency, E 131
jurisdiction 13

knuckle radius 138

ladders and walkways 85
leak test 29
lined vessels 78
low-life cap 39

magnetic particle testing 281
magnetic testing (MT) examination techniques 275
  prod technique 275
  yoke technique 276
material choice 127
material verification (PMI) technique 287
MAWP (maximum allowable working pressure) 15, 21, 46, 109–110, 130, 143, 144–145, 286
  calculations 48
MDMT (minimum design metal temperature) 16, 28, 209, 288
mechanical fatigue 93
mechanical tests 235
metal inert gas welding (GMAW) 214
  consumables 219
minimum specified RT/UT requirements 177–179

NDE of repair welds 185
NDE qualifications 18
NDE requirements
  API 510 257–261
  ASME V 262–283

Index

new construction 6–7
newly installed vessel 46
nominal thickness 179, 196
non-essential variables 231, 241
non-metallic liners 79
notch-toughness tests 236
nozzle compensation 162–167
nozzle design 161–167
nozzle reinforcing limits 167–168

on-stream inspection 32, 34, 82, 288
on-stream pop test 119
outline repairs 71
over pressure 110

penetrameters 263
penetrant testing (PT) examination techniques 270
permanent repair 69
pilot operated pressure-relief valves 112–113
pitting interpretation 52
pneumatic test 150–152
  procedure 150
P-numbers 237, 242
post-weld heat treatment (PWHT) 194
  API 510 overrides 197–201
  replacement by preheat 198, 200
  temperatures and holding times 195

pressure relief valve 107
  pilot operated 112–113
pressure testing 27, 67, 143, 146–148
pressure vessels
  materials of construction 78–79
  types of 77–78
pressure-relieving devices (PRVs) 39, 41
  handling 122
  inspection and testing 119
  seat lapping 117
  terminology 107
Procedure Qualification Record (PQR) 225, 234, 245
  format 229–230
  validation 255
PRVs 41
  inspection interval 41
  inspection periods 40
  seat lapping 117
  terminology 107

qualify the welder 226
qualify the WPS 226

radiographic density 266
radiographs, evaluation 267
ratio of material stress values 147, 150
ratio of stress values 146
RBI assessment 287
RBI study 36
reinforcement pads 183

reinforcing pad 166
relief valve 107, 110
remaining life 285
repair organization 15
repair techniques 68–71
repair, alteration, re-rating 61–74
repairs 17
replacement of PWHT by preheat 198, 200
replacement rule, 2 for 1 192
required thickness, t_{min} 144
re-rating 16, 62–68
  flowchart 66
  reasons for re-rating a vessel 64
  which code edition? 65
re-test of rejected welds 191
risk-based inspection (RBI) 23, 24, 37
RT grades 127–128
  of UG-116 129
  terminology 128
RT levels 177
RT requirements
  of ASME VIII 185
  of UCS-57 187
RT-1 128
RT-2 128
RT-3 128
RT-4 128
rupture disc 108
rupture disc devices 114–115

composite rupture disc 115
forward acting 114
reverse acting 114

safety relief valve 107
safety valve 107
seamless vessel sections 176
self-reinforcement 182
set-in nozzle 182
set-on nozzle 182
shielded metal arc welding (SMAW) 213
  consumables 218
short life cap 38
similar service 38, 86
S-numbers 242, 284
sodium hydroxide (NaOH) 101
sour service 70
spherical radius 139
spot RT 186
  of welded joints 189–192
Standard Weld Procedure Specification (SWPS) 231
standardization blocks 279
static head 145
stress corrosion cracking (SCC) 100
stress ratio reduction 209
stress
  axial (longitudinal) 132
  hoop (circumferential) 130
submerged arc welding (SAW) 214, 216

Index

consumables 220
sulphidation 287
sulphidation corrosion 99
supplementary variables 231, 241

tapered transitions 179
technical committees 2
TEMA (Tubular Exchangers Manufacturers Association) 80
temper embrittlement 103
temper-bead welding 198
temporary repairs 69, 72, 284
tension tests 235
test gauges 149, 152
test pressure 28
thermal fatigue 92
thermal relief valve 111
thickness meter 280
torispherical head 137–138
  geometry 138
transition temperature 16, 28
$t_{required}$ 44
trevitesting 119
tungsten electrode 214
tungsten inert gas welding (GTAW) 214
  consumables 219

UCS-56 table notes 195
UCS-66 steps 205–208
units 4
UW-12 joint efficiencies 175–177

vacuum valve 108
vessel head shapes 134
vessel repairs 63
vessels for lethal service 174
voluntary heat treatment temperature reduction 209

weld overlay repairs 70
weld preheating 199
Weld Procedure Specification (WPS) 225, 234, 245
  format 227–228
welded joint categories 172–173
welded joints, design of 179–181
Welder Performance Qualification (WPQ) 225
welder qualification 234, 239
welding
  API RP 577 Welding Inspection and Metallurgy 212
  consumables 215, 217
  flux-cored arc welding (FCAW) 215, 217
  metal inert gas welding (GMAW) 214
  metal inert gas welding (GMAW), consumables 219
  qualifications and ASME IX 225–256

shielded metal arc welding (SMAW)   213
shielded metal arc welding (SMAW), consumables   218
submerged arc welding (SAW)   214, 216
submerged arc welding (SAW), consumables   220
tungsten electrode   214
tungsten inert gas welding (GTAW)   214
tungsten inert gas welding (GTAW), consumables   219
welding consumables   215, 217
wet H$_2$S damage   102
  hydrogen blistering   102
  hydrogen induced cracking (HIC)   102
  stress oriented hydrogen induced cracking (SOHIC)   102
  sulphide stress corrosion cracking (SCC)   102